当代建筑师系列

朱小地
ZHU XIAODI

朱小地编著

中国建筑工业出版社

图书在版编目(CIP)数据

朱小地/朱小地编著. —北京：中国建筑工业出版社，2012.7
(当代建筑师系列)
ISBN 978-7-112-14392-4

Ⅰ.①朱… Ⅱ.①朱… Ⅲ.①建筑设计-作品集-中国-现代
②建筑艺术-作品-评论-中国-现代 Ⅳ.① TU206 ② TU-862

中国版本图书馆 CIP 数据核字（2012）第 115456 号

整体策划：陆新之
责任编辑：刘 丹 徐 冉
责任设计：赵明霞
责任校对：张 颖 刘 钰

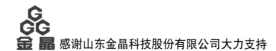
感谢山东金晶科技股份有限公司大力支持

当代建筑师系列
朱小地
朱小地编著
*
中国建筑工业出版社出版、发行（北京西郊百万庄）
各地新华书店、建筑书店经销
北京嘉泰利德公司制版
北京顺诚彩色印刷有限公司印刷
*
开本：965×1270毫米 1/16 印张：9¾ 字数：272千字
2012 年 8 月第一版 2012 年 8 月第一次印刷
定价：98.00 元
ISBN 978-7-112-14392-4
　　　（22463）

版权所有　翻印必究
如有印装质量问题，可寄本社退换
（邮政编码 100037）

目 录   Contents

朱小地印象	4	Portrait
"山水楼台"会所	8	View House
中国石油天然气集团公司总部大厦	32	China Petroleum Headquarter Building
奥林匹克公园中心区规划与设计	44	Planning and Design of the Olympic Park
"秀"吧	52	Xiu Bar
哈德门饭店	76	The Reconstruction of Hademen Hotel
"旬"会所	84	Xun Club
西安唐大明宫国家遗址公园御道广场	116	Emperor's Way Square of Xi'an Daming Palace National Heritage Park
朱小地访谈	146	Interview
作品年表	150	Chronology of Works
朱小地简介	155	Profile

朱小地印象

文／黄元炤

朱小地，1964年出生，1988年毕业于清华大学建筑系后，进入北京市建筑设计研究院工作，两年后被派往海南分院，1994年回北京工作至今。2002年他担任院长以来，一直致力于推动设计院的改革，使传统的事业单位走向市场化发展的道路。他提出并实施的设计院工作室模式极大地调动了设计人员的积极性，并在业内产生深远的影响。在繁忙的管理工作之余，他仍坚持建筑创作，寻找属于他自己的设计方法和设计个性，并不断通过建筑实践表达着自己对世界与人生的领悟。

在20世纪90年代初期，经历了当时下南方的建设大潮，他在海南分院期间参与了一些项目，其中代表性作品就是海口寰岛泰得大酒店，他负责建筑专业设计的全部工作。这个项目在满足酒店基本功能需求的基础上，空间设计试图营造出一种热带的度假氛围，建筑内部设计有椰树、海景观光电梯，33米高的四棱锥网架结构大堂，提供给室内充足的光线与亮度，所以，可以观察到不管是植物、光线的应用，抑或空间的营造，都是贴近于当地性，充分运用当地的自然资源，给予建筑一个生成的机会，是在功能考量的基础上，偏向于当地性（气候、环境、文化）设计思考的地域倾向。朱小地认为他只希望能设计一个场所，营造一种氛围，然后当人进入的时候，能够感受到这一切。

氛围，已成为朱小地设计思考和系列作品的一项重要特征，而氛围就是一种空间气氛与情调的营造。在北京银泰中心顶层的"秀"吧与"旬"会所这两个项目中，朱小地运用了水、树木、植栽、音乐、灯光，试图塑造出时尚、休闲、轻松的氛围及中国特色的意境，企图将深远的中国文化体现在现代空间之中，让人的视觉与感受进入一场仪式性的流程，用氛围去影响人的行为与观感，这两个项目又因各自所处的周边环境与命题条件，而有不同的设计与氛围的表述。

"秀"吧，以高傲的姿态置生于高层建筑群房的顶层，平面布局是在一个限定的范围内做功能性的考量。而这个功能性，朱小地运用了中国传统序列的布局方式，将建筑处于一进一落、一虚一实的状态。另外，他将宋式屋顶置于建筑空间之上，仿佛是传统宅第建筑的现代再现，制造出在现代时尚的情调下又有点中国古典的氛围涌动。而宋式大屋顶之下又是玻璃与混凝土的现代形制与材料，体现出现代与传统在建筑、空间、材料与氛围中的彼此依存与融合。他想在传统建筑形态中体现现代时尚的氛围，这样的意图十分明显，设计结果虽然处于现代，却是偏向于传统、地域的思考，也带有点折中的味道。

"旬"会所，是个改建与加建作品，是将新的体量寄生与附加在旧的系统上。在设计中，大门是不让直接进去的，人要往两边走进去，这是一个中国式的先抑后扬的缓冲地带。然后朱小地用水、植物与步道隔开建筑内外之间的关系，同时用墙与廊组合起来的片段，将人的视觉与动作置于一转一折、一停一留之间，创造出一种转折或是端景式的路径。当人一转一停后可以突然看到一个景，比如户外吧台、四人座桌椅，或是围塑出的庭园，然后再由游廊进入到内部空间，所以，朱小地把传统空间的游园移动过程表达出来，在路径设定上偏向于抽象性的隐喻，而玻璃与钢等现代材料也附加在旧有砖墙上，从旧与新的融合中塑造出现代时尚的氛围。

朱小地速写
作者：高 冬（画家）

散点，是朱小地常用的设计概念，他想让人能够来走他设计的心灵道路，而不是去追求形式的夸张。在"秀"吧与"旬"会所中，朱小地所表现的形式语言中规中矩，各说各话，并不突兀。另外，由于这两个项目，一个是进、落、虚、实的传统序列的布局方式，一个是转、折、停、留的传统游园的布局方式，朱小地又将项目中各种大小物件散点地布置于设计中，用散点去打破一个整体，使得建筑没有内外之分，空间挣脱束缚，介于退隐与自由之间。他试图让建筑表达出一种精简，一种沉寂与宁静。散点，朱小地的理解就是编辑层次的概念，从不同的方向与角度去考虑设计，需渐进式的体验，从中反复寻找转合的可能性及相互如何对应的关系，激生出设计中的亮点后把它们串联起来。

传统与现代，这两者之间的冲突与结合常出现在朱小地的设计当中。他的理解是将时间作为表述设计的一个轴线，而时间轴的概念就是期望用轻松与简单的方法，创造停留的意义，将设计中的各种因素找到一个重新集合的机会。"秀"吧，是一个传统与现代的时间轴体现，当看到步移景异的节点与场景与宋式大屋顶相结合，人的思绪被拉回到久远古老的年代，仿佛游走在宫城大街上或是穿行在大院宅第之间，是很传统的；而到了晚上，音乐、灯光、舞曲、美酒、佳肴与庭院休闲座椅，人的视觉、听觉与味觉又都处于现实生活的时尚氛围中。所以，一瞬间，人的思绪与感觉是来回在千年之间，有时怅然若失，有时又微醺迷离，传统与现代的冲突感一览无遗。

文化传统，他认为中国传统建筑包含的内容极其丰富，但归纳起来可以有具体的几个明确特征。

首先是方位，也包括轴线对位、左右对称，这是中国传统建筑空间营造的根本。然后是院落，他认为中国传统建筑的精髓在于院落，体现的是人与建筑的对应关系，一种从低到高的对应关系，这会使人很快意识到自己所处的位置，处于一个什么样的关系之中。最后是层次，是建立在庭院与建筑之间多重的空间演绎，这是中国传统建筑的意境。根据以上对传统建筑与人之间互立、互含的二元关系，没有必要特别强调某一方，设计中尽量逃避视觉醒目的表现形式。比如SOHO现代城、中国石油天燃气集团公司总部大厦、哈德门饭店重建方案等。在形式放松的基础上，朱小地在他所有的项目中都赋予传统院落的现代再现，不管是平面的还是立体的，如SOHO现代城的立体四合院，哈德门饭店重建方案中的空中四合院。从朱小地的这些设计思考与观点，可以观察到他的设计作品都与传统文化保持着密切的关联，即使建筑表现得简单、现代，但思想上却是偏向于传统、地域的设计倾向，他的态度其实非常明显，表面是一个干净的现代躯壳，内在却荡漾着传统的文化。

"层"论，是朱小地在基础理论方面研究的课题，并作为指导自己的建筑实践和进行建筑评论的重要依据之一。他认为当代建筑越来越依靠工业化生产的方式、由品种繁多的材料制造的构件建筑起来，并以彼此相类似的形式在城市中形成建筑的聚集，导致建筑的形式限定的内与外已很难界定。在建筑密集的城市环境中，被多重层状界面划分构成的空间已经逐渐成为人们感知城市的基本印象。因此，必须建立与之相适应的理论体系才能正确地开展设计与评论。笔者通过剖析当代城市中典型建筑的设计方法，提出广泛存在于建筑形式中的"层"的概念，明确"层"是当代建筑的基本词汇，"编辑"则是对于"层"的理性处理方法，进而将"层"的概念扩展到城市，认为城市空间是"层组"的多维集合，以此建立认识建筑的新的思维框架。建筑的个性化创作尽管存在着不同的途径，但可以归纳为对形成围合的"层组"进行独特的秩序表达，并通过"编辑"的方法形成建筑空间形态的过程。

动态规划，朱小地认为建筑师一定要将专业领域扩展到城市空间和传统空间的研究，并从中确立设计的方向。在城市空间研究方面，他关注世界城市的概念，包括城市比较学与城市经验的获取，然后提出动态规划理论，强调公众利益对应的是公共空间，以自由开放的步行系统为标志。因此，城市规划必须就城市公共空间的范围与权责界定作出回应，这是中国当今城市规划建设与经济发展的必然要求。动态规划理论明确界定了城市空间与建设项目的利益关系，形成建设方自我约束的机制，确保城市空间的规模和质量；平等对待城市中每个建设项目，充分发挥土地资源利用价值；激活社会力量共同参与建设任务的可能性、主动性和创造性，使城市规划与建设处于动态发展的状态，保持城市活力。

综观朱小地的设计思想与作品，他是站在传统建筑文化的思考点，期望通过现代的方式将传统建筑文化的价值表现出来，这不是复古，而是一种新的尝试与诠释，对应的是"传统如何走向现代"的历史命题。另外，他又把建筑定位在城市空间尺度的出发点上，找到适宜于人的尺度与环境，然后再从环境的角度切入设计，运用研究成果与手法处理空间，最后塑造出一个场景与氛围，让人们从中去感受，这才是他理解的建筑设计。

Portrait

By Huang Yuanzhao

朱小地速写
作者：刘克成（建筑师）

Zhu Xiaodi, born in 1964, after graduating from the School of Architecture, Tsinghua University in 1988, went to work at the BIAD (Beijing Institute of Architectural Design). After he was appointed as the president of the BIAD, he devoted himself to reforming its system and pushing forward the establishment of local branches and studios. In addition to undertaking the task of reformation, he persisted on architecture designs, seeking for design methods, concepts and values of his own through trials and errors.

Zhu went through the tide of going south in the architecture sector, which was in vogue in 1990s. He took part in some projects in the BIAD Hainan Branch, of which a representative work was Huandao Tide Hotel, Haikou. He himself was in full charge of it, and engaged in all its architecture-related tasks. It is a design that applies plants to buildings, just in consistent with the tropical phenomenon of numerous plants in local environment. Besides, in view of the sufficient sunlight in Hainan throughout the entire year and its high photosynthetic capacity, a 27-meter-high hall of pyramid space truss structure is designed to let in plenty of light and illumination. Therefore, it can be seen that it is a design, closely based on the locality, in terms of plants, application of light and creation of space. The triangle turriform spire against the blue sky has an abstract symbolic significance. It's like a bell tower of west classical church or a Chinese classical tower. Zhu thinks that he merely hopes to design a place where people in it can feel the atmosphere.

The atmosphere constantly appears in his design. Creating an atmosphere has been a key point in his thought of design. When he took part in designing Xiu Bar located at the top of Beijing Yintai Centre, Zhu made use of water, trees, plants, light, lamp and music to create a kind of fashionable and relaxing atmosphere, trying to involve the Chinese oriental culture into the modern space. He wanted people to deeply involve themselves into his designs as much as possible, with their vision and sense amazed by the atmosphere, which is a process like a ritual.

The Xiu Bar, located at the top of the building, is a kind of parasitism and annex. The graphic layout is of a functional consideration in a limited area. Zhu uses the layout of Chinese traditional sequence. Moreover, the curved roof, a typical roof of the Song Dynasty, is attached to the top of buildings. It seems that the ancient government or ministry councilor's residence was reproduced in a modern style, with fabrication of Chinese classical atmosphere under the sentiment of modern vogue. The curved roof is constructed with the modern glass and concrete, which embodies the coexistence and combination of modernity and tradition in terms of structure, space, material and atmosphere. We can see his intention very clearly that he wants to express a modern and fashionable atmosphere in a traditional building form. Though the outcome of this design exists in the modern world, it's actually a more traditional and more local thought, which is more or less eclectic.

The Xun Bar is a project of rebuilding and extra building, that is, parasitizing and attaching a new structure to the old one. In design, people are not allowed to enter directly through the portal but to walk in from the two side corridor entrances; thus, the portal serves as some kind of buffer. And then Zhu utilizes water, plants and paths to disconnect the buildings' interior and exterior parts. Meanwhile, the assembly composed of walls and galleries directs people's sights and actions to twist and turn, stop and linger once in a while, creating a kind of turning or scenic path. So Zhu combines the paired-and-balanced conception of the courtyard with the traditional space movement of garden-visiting. He prefers using abstract metaphor in the design of path. And modern materials like glass and steel are added to the original brick wall. That combination of old and new factors is to create a modern fashion atmosphere.

Scattering is a design concept that Zhu commonly uses. He intends to let people walk on his soul path, rather than to pursue an exaggeration of the form. In the design of the Xiu Bar and the Xun Club, the stylistic dialogue that he demonstrates is appropriate. Each of them expresses its own meaning, which seems not abrupt at all. Zhu attempts to make buildings with a simple, still and silent sense. He understands scattering as the concept of arranging structure levels which means to consider design in different aspects. It needs gradual experience, searching for the steering possibility repeatedly and the correlations of how to correspond, and then arouse the bright spots of the design, line them up, make choices and finally file one thing.

We can always find the conflict and bond between the tradition and the modern, the new and the old, in Zhu Xiaodi's design. In his view, the time axis is used to translate the design. And the utilization of time axis can offer a chance to recollect all different elements in the design. The Xiu Bar is a time axis demonstration of the tradition and the modern. When seeing nodes and scenes designed in a way of "sceneries changing after step moves", with the curved roof in the Song Dynasty style, we will be drawn back to an ancient time long ago. It would seem that, in the day we wander along the palace avenues or pass through residence with yards in a traditional atmosphere. While in the evening when the music and lights are on, dance music in the bar, wine fumes and courtyard leisure chairs, all we see, hear and taste are immersed into an atmosphere of modern fashion, which is quite contemporary. Therefore in a moment, people seems to travel back and forth in time over a thousand years. Some may feel lost while others may be slightly blurred. The conflict of the tradition and the modern is in plain sight, which serves as a display and transition in the time axis that Zhu applies in his design.

As for cultural traditions, Zhu considers that Chinese traditional architecture is extremely abundant in contents, however with several specific characteristics by summarizing. First is direction, also including axis symmetry and bilateral symmetry, which is the foundation for creating Chinese traditional architectural space. Second is courtyard. He takes courtyard as the essence of Chinese traditional architecture, which reflects the correspondence between human and architecture. Such low-to-high correspondence enables people to realize where he is and what kind of relationship one is in. The last one is level, which is the multiple spatial illation established between courtyard and architecture and also the artistic conception of Chinese traditional architecture. According to the above mutually independent and also included binary relations between people and traditional architecture, there is no architectural meaning in case of no people and also no need for specially emphasizing a certain party, so his design tries to avoid eye-catching formal visual expression, such as SOHO Modern City, China Petroleum Plaza, reconstruction project of Hademen Hotel, etc.. Based on form relaxation, Zhu Xiaodi endows traditional courtyard with modern representation in all his projects, either planar or three-dimensional, such as three-dimensional quadrangle in SOHO Modern City and air quadrangle in reconstruction project of Hademen Hotel. It can be observed from these design thinking and viewpoints of Zhu Xiaodi that all his design works bear close relations with traditional culture. Even simple and modern architecture tends to be traditional and regional design inwardly. Actually, his attitude is rather clear: it is clean and modern in outer form, however with traditional culture rippling inward.

"Level" theory is Zhu's research topic in basic theory, also regarded as one of the important basis for his own architectural practice and architectural criticism. In his opinion, as current architecture increasingly relies on industrialized production and is built up by construction members produced by various materials, and also forms architecture cluster in mutually similar forms, it is of certain difficulty in defining inside and outside defined by forms of architecture. Under urban environment of dense architecture, space formed by multi-layer interface division has gradually become the basic impression of people perceiving a city. Therefore, corresponding theoretical system must be established to conduct design and criticism correctly. Through analyzing the design methods of current typical architecture, the author puts forward "level" concept broadly existing in architectural forms, and clarify "level" as the basic word of current architecture and "compilation" as the method of rationally processing "level". He further expands the concept of "level" to city and considers city space as multidimensional set of "level group", so as to establish new thinking framework of recognizing architecture. Although individualized creation of architecture bears different ways, it can be summarized as the process of performing special order expression for "level group" that forms enclosure, and forming architecture space appearance through "compilation" method.

As for dynamic planning, Zhu Xiaodi thinks that architect must expand professional field to the research on city space and traditional space, and determine the direction of design therein. From the aspect of researching city space, he pays attention to the concept of world city, comparing study of city and acquisition of urban experience, and then puts forward the theory of dynamic planning and emphasizes the correspondence of public space to public interest, with free and open pedestrian system as the sign. Therefore, urban planning must give response to the scope and responsibility defining of urban public space, which is the inevitable requirement of current urban planning and economic development in China. In summary, the theory of dynamic planning clearly defines the benefit-based relationship between city space and construction project, with self-restraint mechanism of construction unit formed to ensure the scale and quality of city space; it equally treats all construction projects in city and gives full play to the value in land resource utilization; it also activates the possibility, enthusiasm and creativity of societal forces to jointly participate in construction tasks, thus locating city planning and construction in a state of dynamic development and maintaining urban vitality.

Of all Zhu Xiaodi's design belief and works, he expects to, based on the traditional architectural culture, perform its value by modern ways. This is not a kind of retro, but a new try and annotation corresponded to the historic proposition of how the architecture moves from tradition toward modern. He believes that he has a responsibility to undertake the proposition. Furthermore, he positions buildings on the starting-point of urban space scales to find suitable environment for people. Then he designs according to the environment, handling the space by research findings and techniques, and finally creates a scene and vibe for people to feel. This is his understanding of architectural design.

"山水楼台"会所 北京
View House, Beijing
2002~2003

建设机构 / Construction Organization：华熙房地产开发有限公司 / Huaxi Real Estate Development Co., Ltd.
项目地点 / Location：北京怀柔 / Huairou, Beijing
建筑面积 / Floor Area：5500m²
建筑高度 / Building Height：15m
设计时间 / Design：2002年4月 / Apr., 2002
竣工时间 / Completion：2003年8月 / Aug., 2003

项目场地位于京郊怀柔区神堂峪风景区，山间公路在西侧，与场地隔河相望。场地的南、东、北三面被山体围合，有堪舆学中所描述的"负阴抱阳"之形态。山脊上一段完整的长城城墙将山体的轮廓线勾勒得更加清晰和秀美，也成为这块用地天然的屏障。建设场地约有几十亩的样子，位于不同高度的几个台地上，建有几栋简易的房子，是以前的业主留下的。

河道在这里较其他地方显得更加宽阔，并形成较深的河床，两侧的陡坡有3~5米的样子。河水在这里还有1米多高的落差。从公路到达对面的用地，必须经过一座石头砌筑而成的小桥，东西走向，横跨在河道当中。

道路的左侧、也就是西侧的山体，从远处的最高峰一直延伸到道路的边缘，形成向东的走势，其曲折的山脊上依然是一段古老的长城，几个烽火台错落有致地立于不同高度的山顶之上。沿山路爬上西侧山坡的第一个烽火台，可以俯瞰对面半坡上的建设场地所处的环境，只见山环水抱、绿荫叠翠。在此观赏云影在山腹间移动的样子，恍惚感觉到时光的味道，令人陶醉其中。

设计从上述最具价值的景观因素出发，将建筑体量分解，按照空间功能将其化为私密、半公共、公共三个独立的建筑。结合台地地形采用错层布置方法，重复的韵律勾描出清晰的空间关系。三栋建筑以西侧的烽火台为向心点，在正南至南偏东20°的最佳采光范围内呈扇形错动分布，最大限度地保护原有地貌并使建筑融入环境之中，也为室内各个空间提供最大可能性的视野。三栋建筑之间用通廊连缀，自然围合出两个内花园，使每栋建筑可同时拥有郊野和庭院不同的趣致。

每栋建筑北侧的楼梯间稍高，成为南侧单坡屋顶的依靠。三层西侧退后，留出朝向水面的3个观景露台，作为室外的活动场地。北栋西侧的大会议室通过8米的悬挑避让岸边的一棵柳树，伸向水面的部分无论在夜晚还是白天，都成为河边的一道风景。

为创造最佳的观赏效果，避免视线的阻滞，设计采用无边框的落地玻璃窗，整片玻璃直接嵌入清

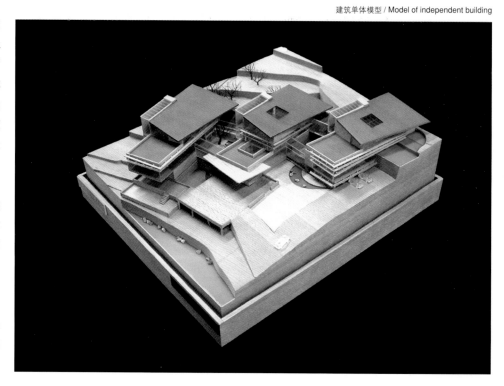
建筑单体模型 / Model of independent building

场地原状 / Original site condition

北栋向水面的大悬挑 / North building cantilevers toward waterfront

水混凝土梁柱内。在南、西两侧外窗之上设置水平木线条遮阳，挡去多余的阳光。庭院内通过流水、静池的设计提供亲切的院落环境，落地窗将庭院内的光影变化引入室内，形成亲切自然的氛围。朝向北侧、西侧的小房间，设计了特定位置的方窗，如画框般提示着窗外的风景。

建筑外墙选用天然和人工两类材料，突出传统与现代、自然与人工的反差与结合。毛石墙贴近自然的环境，有从环境中生长的感觉。屋面采用青石片，是当地民居的一种习惯做法。遮阳的松木条板，强化了建筑与环境的和谐。清水混凝土、低反射玻璃在色彩、质感等方面更加突出现代技术的成果。设计将这些材料的使用延续到室内设计当中，所有公共空间的顶棚、楼梯间墙面均采用了清水混凝土，并多以透明的玻璃灯具为主，凸显清水混凝土的素雅。围绕内庭园设置了部分毛石墙，在会客区设计了毛石砌筑的壁炉通至屋顶。通透的低反射玻璃则降低了建筑外墙的界限，将室外不可多得的自然景色引入室内。

各栋的楼梯间的顶层屋面、顶层卫生间屋面、小餐室的屋面、中栋庭院的屋面都设计了可开启的天窗，提供自然通风条件。室内地面采用青石板、木地板和白色大理石，控制室内的整体色调，力求使人的视线更多地在自然景致上驻留。

The project is located at Shentang valley scenic spot at Huairou District, Beijing, with intermountain road in the west across the river. The south, east and north of the site are surrounded by mountains, which form the pattern of "shady well combined with sunny side" described in geomancy. A complete section of Great Wall on the ridge draws the outline of the mountain vividly and elegantly, which also serves as the natural barrier for this area. The construction site, covering an area of about dozens of mu, is located at several platforms with different heights and left with several simple houses constructed by the former owners.

The riverway here is broader than that at other locations and forms deeper riverbed with slope of 3 to 5 meters at both sides. Water head more than one meter is also formed here. The only way from the road to the opposite site is an east-west stone bridge spanning across the riverway.

The left side of the road, i.e. mountain at the west side, extends to the road edge from the peak far away, forming the eastward trend. On the tortuous ridge of the mountain is a section of ancient Great Wall with several well-spaced beacon towers scattered at the peaks with different heights. After climbing up the west slope along the road, one can arrive at the first beacon tower to overlook the opposite construction site at hemi slope with mountain surrounded, river girdled and green overlapped. Watching the cloud shadow floating between mountainsides, a taste of time seemingly appears in a trance, enchanting and intoxicating.

The design based on the most valuable landscape elements abovementioned decomposes the building dimensions and divides it into three independent buildings according to spatial functions, namely private, semi-public and public areas. Split-level layout is applied in combination with the platform terrain and the repeating rhyme outlines the distinct spatial relationship. Taking the beacon tower in the west as core, the three buildings are distributed in the form of a fan within the best lighting spectrum in the south and at 20 degree from the south to east, thus protecting the original landform to the greatest extent, integrating the buildings into the surroundings and also providing the maximum view for each interior space. The three buildings are connected and clustered with vestibules, naturally establishing two interior gardens and simultaneously leaving each building with both suburban and yard funs.

The staircase at north side of each building is slightly high, on which the single slope roofing at south side relies. The west side on the third floor is backward, leaving three viewing terraces facing the water to serve as the outdoor activity space. The large meeting room at the west side of north building is overhung for 8 meters to keep away from a willow at river bank, and the part extruding the water becomes a riverbank scene no matter at night or in daytime.

For the purpose of creating the best aesthetic effect and avoiding the visual retardation, the design applies frameless landing glass window with the whole piece of glass directly embedded into cast finish concrete beam column. The exterior windows at south and west side are equipped with horizontal wooden linear sun shadings to keep off the superfluous sunshine. The flowing water and static pond in the yard provide amiable courtyard environment and the French window introduces the changing shadow in the yard to the room, forming a harmonious atmosphere. Square windows are provided in specific positions in small rooms facing north and west sides, presenting the outdoor scenery like a picture frame.

The buildings' outer walls apply both natural and artificial materials to highlight the contrast and combination between tradition & modernism and nature & artificiality. The rough wall is close natural environment, leaving people a feeling of growing from surroundings. The roof applies bluestone pieces, a local common practice. The adumbral pine batten strengthens the harmony between buildings and environment. The cast finish concrete and low reflecting glass further highlight the achievements of modern technology in terms of color and texture, etc.. The usage of these materials extends to interior design. All public roofs and staircase wall surfaces apply cast finish concrete, with transparent glass luminaries as main decorations to express a sense of freshness and elegance of cast finish concrete. Partial rough wall is constructed surrounding the interior yard. A fireplace constructed with rough stone and connected to the roof is designed at the meeting area. The transparent low reflecting glass minimizes the boundary limits of the outer walls and introduces the precious natural scenery into room.

The top roofing of each staircase, top washroom roofing, dinette roofing and roofing of the middle building yard are all designed with skylights which can be opened to provide natural ventilation. Interior floor applies bluestone, timber floor and white marbles to control the overall indoor color tone, with an intention to draw on visitors' attention to the natural scenery for a longer time.

1 北栋
2 中栋
3 南栋
4 配套用房
5 内院
6 竹丛
7 溪水（卵石岸）
8 草坪
9 水池
10 主入口平台
11 木平台
12 鱼池
13 山溪
14 桥
15 主入口大门
16 小水潭（卵石岸）
17 草坪
18 果树林

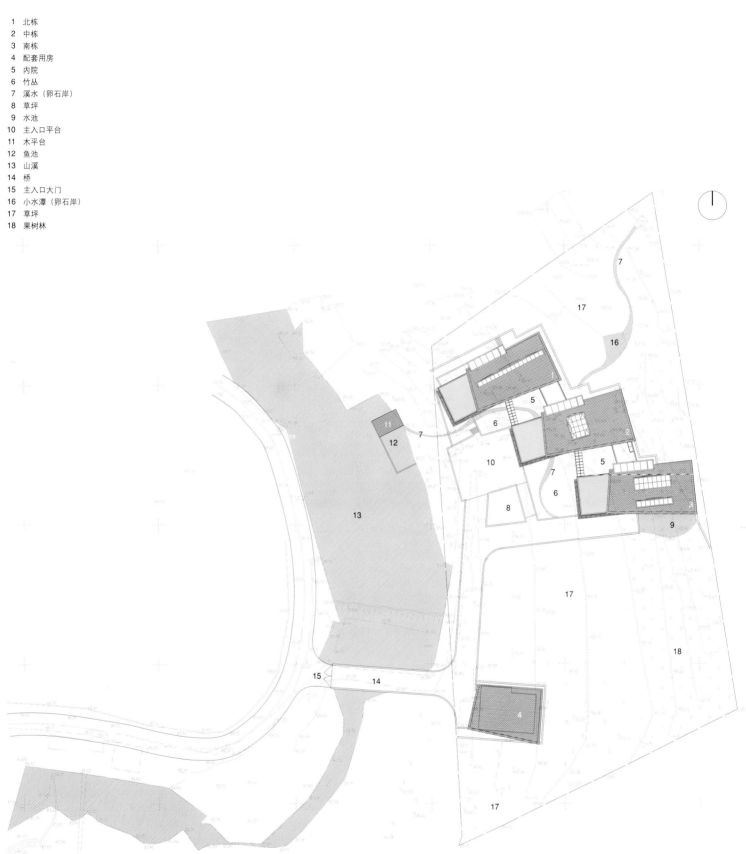

总平面 / Master plan

1	中式主卧
2	西式主卧
3	日光室
4	更衣室
5	早餐间
6	中式顶采光化妆间
7	西式顶采光化妆间
8	西式书房
9	卧室
10	起居室
11	大客厅上空
12	客房
13	库房
14	和室
15	露台
16	屋面
17	玻璃屋顶

三层平面图 / Plan of the 3rd floor

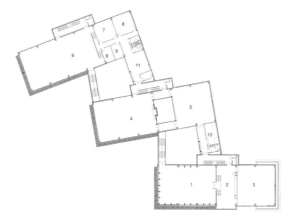

1	中式书房
2	小客厅
3	小过厅
4	大客厅
5	视听室
6	多功能厅（大餐厅）
7	西式小餐厅
8	中式小餐厅
9	棋牌
10	小餐厅
11	台球厅

二层平面图 / Plan of the 2nd floor

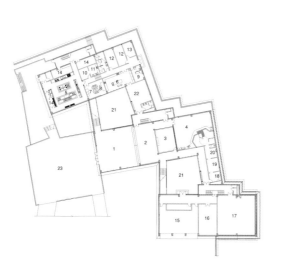

1	门厅
2	雕塑藏品厅
3	水光庭
4	陶吧
5	专业厨房
6	家庭厨房
7	洗衣间
8	熨衣间
9	管家
10	工人餐厅
11	工人服务
12	工人房
13	活动室
14	库房
15	车库
16	酒窖
17	藏品库
18	更衣
19	桑拿
20	按摩
21	内院
22	家庭餐厅
23	平台

一层平面图 / Plan of the 1st floor

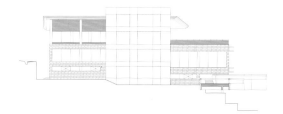

北栋北立面图 / North elevation of the north building

北栋西立面图 / West elevation of the north building

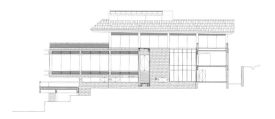

北栋南立面图 / South elevation of the north building

北栋东立面图 / East elevation of the north building

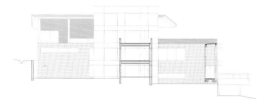

中栋北立面图 / North elevation of the middle building

中栋西立面图 / West elevation of the middle building

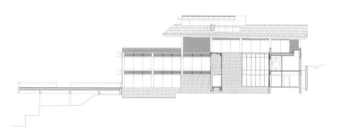

中栋南立面图 / South elevation of the middle building

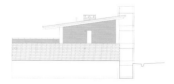

中栋东立面图 / East elevation of the middle building

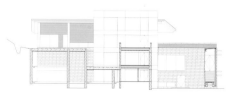

南栋北立面图 / North elevation of the south building

南栋西立面图 / West elevation of the south building

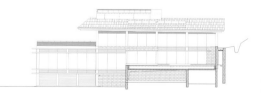

南栋南立面图 / South elevation of the south building

南栋东立面图 / East elevation of the south building

1

2

3

1. 北栋楼梯间南墙立面图
   South elevation of staircase of the north building

2. 中栋楼梯间南墙立面图
   South elevation of staircase of the middle building

3. 南栋楼梯间南墙立面图
   South elevation of staircase of the south building

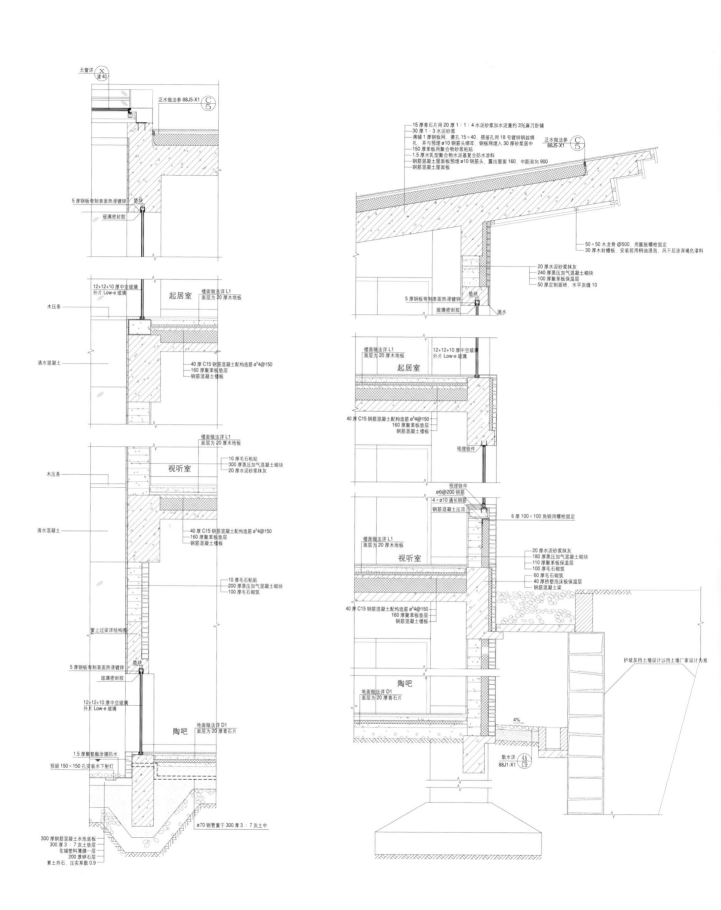

外墙大样 / Details of external wall

中国石油天然气集团公司总部大厦 北京
China Petroleum Headquarter Building, Beijing
2003～2008

建设机构 / Construction Organization：中国石油天然气集团公司 / China National Petroleum Corporation

项目地点 / Location：北京东直门 / Dong Zhimen, Beijing

建筑面积 / Floor Area：200800m²

建筑高度 / Building Height：90m

设计时间 / Design：2003年10月 / Oct., 2003

竣工时间 / Completion：2008年8月 / Aug., 2008

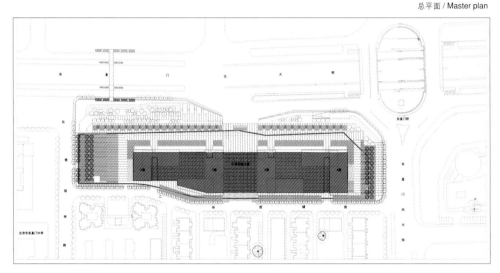

总平面 / Master plan

中国石油大厦设计方案正面回应了场地制约的挑战。建筑沿南北方向一字摆开，恰好确定了四栋南北向的办公建筑，充分发挥了场地的资源条件。在此基础上，将南北两组建筑在东西方向上加以延长，形成"L"形的平面和两个相对围合的空间，中间的部分再通过上部的联系，形成集团的入口礼仪大厅。分散布局最大限度地满足了建筑主体尽可能多的南北朝向，确保了建筑主体的自然采光和通风，使其形成一组主次分明、错落有致的建筑群。

整栋建筑沿东二环路一字排开，形成长达250米的城市界面。四栋办公建筑之间形成的巨大空间标定了现代城市建设在尺度上发生的巨变，提示着城市空间的存在，引起公众对建筑的关注，在空间上将城市与建筑紧密联系在一起。

在四栋办公建筑平面布局确定之后，建筑裙楼部分被设计成贯通南北的连续空间，将大厦的主要出入口有机地连接起来，在解决功能的同时，创造出连续、宏大的共享空间。其中，大厅长宽约40米，高53米，采用张拉双索玻璃幕墙，将室外景观引入室内。

建筑外立面选材以石材为主，整体风格力求稳重大方。立面肌理采用竖线条的设计方式，使建筑获得高耸挺拔的视觉效果，凸显中石油集团蓬勃向上的企业精神和发展前景。建筑幕墙为石材竖线条间隔玻璃幕墙，在保证大厦总体石材感觉的同时，最大限度地增加了办公空间的通透率，有效地控制光线的反射。

平面柱网采用8.1米×8.1米/5.4米，以1.35米为基本单元，将建筑设计纳入模数体系内，使建筑从内到外，从实体到空间都严格遵循统一的逻辑和秩序，从而体现出建筑严谨的风格，表达结构逻辑和空间秩序。

大厦结构采用型钢混凝土结构，同钢筋混凝土结构相比，可降低结构自重，减小断面尺寸，加快施工进度；同钢结构相比，可降低造价，提供防火性能。由于梁柱断面尺寸的减小，使建筑的使用效率得以提高，吊顶的高度得以增加，从而改善了办公环境。

清晰的设计理念和方法得到了建设方的鼎立支持，也为大厦的工程设计整合各类技术方法、提高完成度创造了条件。在设计过程中，从城市空间到建筑单体，从功能要求到平面布局，从室内环境到室外景观，从细部节点到空间尺度，从模数控制到逻辑关系，从材料工艺到技术体系，从设备配置到系统集成，从构造措施到成本平衡，各专业工种努力做到理性的表达和总体的协调，将整体设计的概念落实贯穿于设计过程中。

生态设计充分考虑大城市生态环境因素的影响，将周边环境与本项目建筑自身之间的不利影响降至最低。中国石油天然气集团公司是以生产能源为主的大型企业，大厦的建设必须充分体现节能意识，尽可能地采用包括先进的新技术、新材料、智能化等多项措施，减少能源消耗，提高能源利用效率，节省业主建设运行费用，将中国石油大厦建设成为具有世界先进水平的绿色、环保、节能的办公大楼。

人性化设计中始终坚持绿色、健康、环保理念，充分考虑中国石油大厦使用者的各种需求，为之提供周全、便捷的人性化服务。要保证建筑物内高舒适度的办公环境，有效控制室内的空气质量、室内的噪声等级以及室内的热环境等，保持空气的清新和洁净，力求营造典雅舒适、安全便捷、尺度宜人并别具特色的场所空间与可持续发展的人居环境。

智能化设计作为中国石油天然气集团公司总部办公场所，其内部功能必须与该企业面向全国、面向世界的形象相匹配。智能化系统是整个建筑的"大脑"，是体现大厦现代化、先进性的重要因素，因此，中国石油大厦必须在高智能、高集成的水准上达到世界先进的智能建筑水平，为大厦的使用者提供一个安全、高效和便利的办公空间。

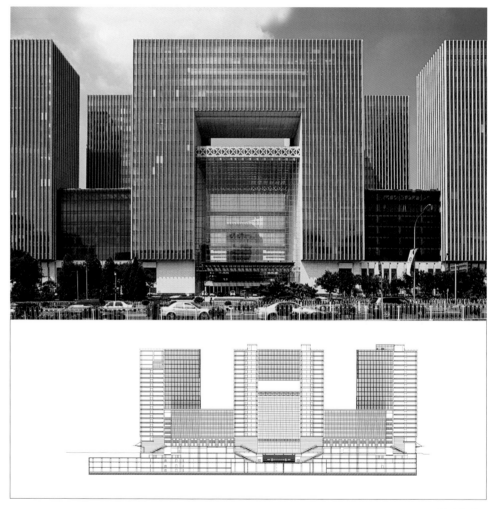

剖面 / Section

The design scheme of China Petroleum Plaza has given a positive response to the challenge of site restriction. The building lines up in "L" shape from south to north, which rightly defines four south-north office buildings and fully utilizes the available space of the construction site. The north and the south ends of the building group extend to the east and west directions respectively, forming two L-shaped structures and relatively enclosed spaces. The middle section is connected to the upper floors and works as the entrance to the grand hall. The dispersed layout furthest satisfies the south-north direction of building body and ensures natural lighting and ventilation, thus resulting in a distinguishing well-arranged architectural complex.

The entire building lines up along East 2nd Ring Road and forms a city interface of 250m long. The large space created by the four office buildings marks the dramatic change of modern urban construction in dimension, which suggests the existence of city space and arouses the public's attention to architecture, thus closely connecting city and architecture in space.

After determining plane layout of four office buildings, annex buildings are designed as continuous space running through south and north. They connect the main passageways of plaza organically and create continuous and grand shared space at the same time of satisfying the functions. Inside the plaza, the hall, about 40m in width and length while 53m in height, adopts stretch-draw double-cable glass curtain wall to bring outdoor landscape inside.

Building elevation selects stone materials and the holistic style is in prudent generosity. Façade texture adopts the design method of vertical moulding, which gives an overwhelming visual effect and highlights the vigorous enterprise spirit and development prospect of CNPC. The building curtain wall adopts stone vertical moulding to space glass curtain wall, which farthest increases the transparency of office space and effectively controls the amount of light reflected at the same time of ensuring the overall stone feeling of plaza.

The area of the plane column grid is 8.1m x 8.1m / 5.4m, with 1.35m as its basic unit. The design of the buildings was brought into modulus system, so the buildings follow strictly unified logics and order from outside to inside and from entity to space. It reveals a rigorous architectural style that indicates structure logic and space order.

The plaza adopts the structure of SRC. Comparing with RC structure, SRC can decrease the dead weight and section size, speed up construction progress, reduce the construction cost and improve fireproof performance. Due to the reduction of the section size of beams and columns, the building's service efficiency gets improved, and the height of suspended ceiling is increased. Thus, the office environment is improved.

Clear design concept and method receive full support from the construction unit and integrate various technical approaches for engineering design of the plaza, thus creating conditions for improving completeness. During design, from city space to individual building, from functional requirement to plane layout, from indoor environment to outdoor landscape, from detail nodes to space scale, from module control to logical relation, from material technology to technical system, from configuration to system integration, and from construction measures to cost balance, all crafts of various specialties try to achieve rational expression and overall coordination and implement the concept of overall design into design process.

Ecological design takes the influence of ecological environment factors of large city into full consideration and minimizes the adverse effects between surrounding environment and buildings of the project. China National Petroleum Corporation is a large-scale enterprise focusing on producing energy, so the construction of the plaza must fully embody the consciousness of energy conservation, adopt advanced new technology, new materials and smart approaches as much as possible, reduce energy consumption, improve energy utilization ratio and save the construction and operation cost of the owner, thus constructing China Petroleum Plaza into an environmental-friendly and energy-saving office building with world-class level.

Throughout the design process, we worked to maintain the concepts of being green, being healthy, and environmental-friendliness. We gave much consideration to the needs of future tenants and worked hard to provide a comprehensive, personalized and convenient service. The engineering guarantees an office environment of high comfort level inside the buildings and keeps tight control of internal air quality, as well as the indoor noise level, thermal environment and so on, with fresh and clean air maintained. This creates a comfortable, safe, convenient, pleasant and distinctive site space as well as sustainable human environment.

As the office place for the headquarters of China National Petroleum Corporation, the built-in functions of smart design must match with the enterprise's image of facing the country and the world. Smart system is the "brain" of the whole building and a symbol of the modernization and advancement of architectural technology present in the building. Therefore, China Petroleum Plaza must reach the world-advanced level of smart building on high intelligence and integrated standards, so as to provide tenant with a safe, efficient and convenient office space.

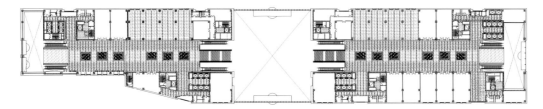

二层平面（地拼）/ Plan of the 2nd Floor

一层平面（地拼）/ Plan of the 1st Floor

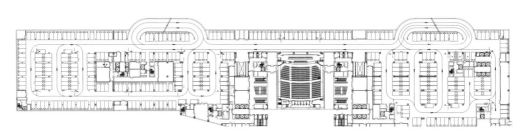

地下二层 / Plan of the -2nd Floor

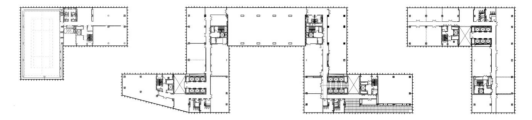

19层平面（地拼）/ Plan of the 19th Floor

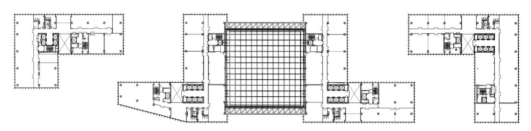

13层平面（地拼）/ Plan of the 13th Floor

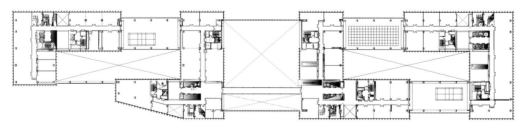

8层平面（地拼）/ Plan of the 8th Floor

奥林匹克公园中心区规划与设计　北京
Planning and Design of the Olympic Park, Beijing
2003～2008

建设机构 / Construction Organization：北京市新奥集团有限公司 / Beijing Xin'ao Group Co.,Ltd.
项目地点 / Location：北京市朝阳区 / Chaoyang District, Beijing
建筑面积 / Floor Area：82hm²
设计时间 / Design：2003年5月 / May., 2003
竣工时间 / Completion：2008年7月 / July., 2008

北京奥林匹克公园位于北京城市中轴线延长线的北端，南北方向跨越北四环、北五环，总长约6千米，总用地1135公顷。包括：北区（森林公园）680公顷，中心区291公顷，南区（原奥林匹克体育中心用地及其南侧预留体育用地）114公顷。

奥林匹克中心区的建设项目包括：8万座席的国家体育场、1.7万座席的国家游泳中心、1.8万座席的国家体育馆、50万平方米会展中心、10万平方米数字北京大厦、51万平方米奥运村运动员公寓，是举办2008年奥运会的核心区域，也是历届奥运会中规模最大的场馆集中区。

中心区规划回避了在中轴线的尽端建设重要项目的设想，主要的体育场馆分别设置在中轴线的两侧，其中国家体育场设置在中轴线的东侧，体型完整且规模宏大；国家游泳馆、国家体育馆和国家会议中心位于中轴线的西侧，与国家体育场取得动态的平衡。这样的处理方式既丰富了北京中轴线的空间形式，又展示了北京中轴线从过去走向未来的历史进程。

中心区景观大道延续北京中轴线，西侧设置平面规整的树阵（绿轴），东侧开挖岸线流畅的"龙形"水系（水轴），使整个中心区形成中轴、绿轴、水轴相互呼应的宏大开放空间。树阵和水系的护佑，柔和了严肃的中轴，强调了中轴空间的亲民性。这与北京旧城的皇城居中、"左祖右社"、等级森严的中轴线形制形成强烈的反差。

方案强调中心区景观整体性，考虑与南区和北区中轴空间的衔接及中心区内部的功能布局，将中心区内部沿中轴景观从南至北划分为庆典广场、下沉庭院、休闲花园三大段落。庆典广场段周围有国家体育场、国家游泳中心、国家体育馆等重要奥运设施，人流量大，赛时各种活动频繁，主要以提供开阔的室外广场环境为主。下沉庭院段以下沉庭院为核心，强化城市空间的立体处理，丰富层次，为赛后城市公众和游客提供交通、休息、娱乐、餐饮、购物等功能。休闲花园作为中心区向森林公园过渡的地带，以人工和自然的景观混合的设计为主，向市民提供开敞的景观空间，同时满足公众集中的休闲娱乐等文化活动。

中轴序列空间以鸟巢、水立方之间的庆典广场为高潮。庆典广场南北长260米，东西宽160米，是奥运会赛时主要庆典活动及大量人流集散地。庆典广场布置了音乐与喷泉相结合的设施，还有LED灯光秀，周边由鸟巢、水立方、树阵围合成热烈、轻松的市民广场。

与中轴相连的城市广场沟通了中轴与两侧重要体育场馆的联系，为各主要场馆展示精美的建筑形

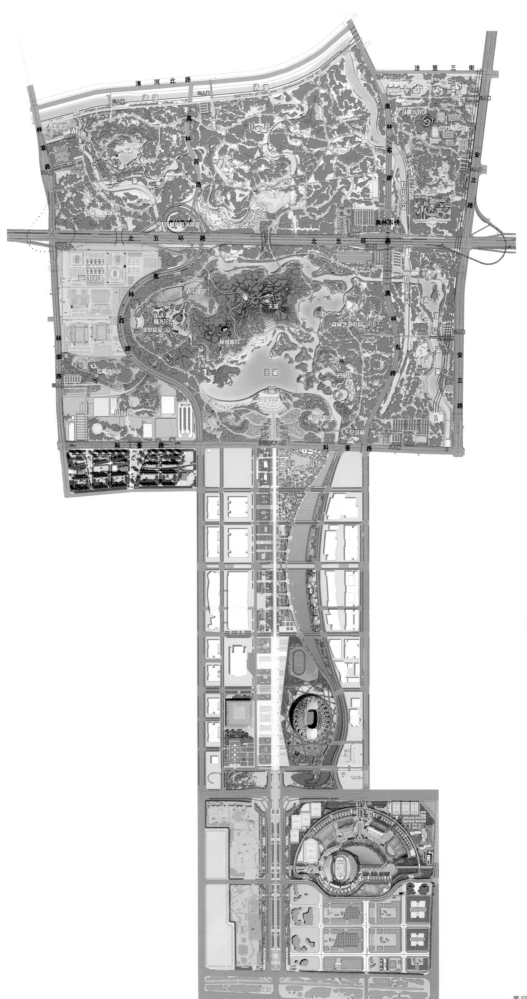

景观总平面 / Landscaping layout

式,提供了必要的疏散和开敞的空间,使独立的场馆以中心区广场为平台形成有机的整体空间环境。

中轴空间的场地设计参考了故宫、天安门广场、天坛、南中轴等地区的中轴线形式,并经各方专家反复论证,最终决定保留古典中轴形式并在设计细节上进行改进,产生了现在的11米中轴线设计。结合项目特点,进行了卫生间、直饮水、信息接入点、自行车停靠、标识系统等服务设施的设计,使中心区景观设计从功能上回归城市公共空间的本质。

位于中心区中部、处于中轴空间与水系之间的下沉庭院,为20万平方米地下商业、地铁出入口以及大容量公共交通的人流集散提供了开敞的出入空间和室外环境,整个中心区景观空间也随之立体化,为人们出行、购物、休闲、会友提供了好去处,

成为新的城市客厅。下沉庭院的增加,使得以中轴为主的地面景观与下沉庭院构成了从地上到地下、从恢宏空间到院落尺度、从大型的广场、建筑到中国式园林的巧妙转换。

下沉庭院由东西方向的城市道路——大屯路分割成两大部分,其下由市政过街通道连接。为方便下沉庭院两侧东西方向的步行联系,在南部下沉庭院之上设置了三个步行廊桥,在北部下沉庭院之上设置了两个步行廊桥,与中轴线垂直,形成北京旧城棋盘格式的交通体系。大屯路和五条步行廊桥将下沉庭院分割成七个近似的院落,在地下形成由南至北的层次递进空间。如同北京传统住宅院落的纵向发展的结构,方正的路网与院落空间的结构体现了中国传统建筑空间意象,至此将超尺度的城市空间降低为大型的院落空间。

建设过程 / Construction progress

整个下沉庭院的设计以"开放的紫禁城"为主题，将紫禁城封闭的红色宫墙断开、错位、联系、贯穿在下沉的七个院落，起到对整个下沉庭院提纲挈领和调整尺度的作用。

1号院截取巨大的宫墙片段置于院落中心，可移动开合的"宫门"采用大尺度LED屏幕，平时向公众敞开；关门时形成舞台背景效果，对面进入下沉庭院的大台阶，又正好利用成为公众驻足观看和休息的露天座位。2、6号院以北京民居四合院为原形，各自表达了对经典的理解和对历史时空的独到见解。同是四合院的原形，产生出完全不同的解析和意境，显示出中国古典建筑元素发展的无限可能。3号院以巨大尺度的鼓墙、编钟、铜笛等古乐器元素，寓意"有朋自远方来，不亦乐乎"。4、5号院作为穿越大屯路的空间，试图尽量减少装饰，使人流在此减少停留的机会，属于过渡的院落。7号院的设计巧妙地结合下沉庭院的坡道、台阶，从中国古典建筑、美术、装饰等元素中汲取片断，力求营造意境，达到了很好的效果。

下沉庭院的传统元素设计给人们带来了强烈的视觉冲击，在细部处理上均有独到之处，内容丰富，寓意深远，令人称奇，是中国建筑史上对传统文化继承与发展的一次崭新而大胆的尝试。

The Olympic Park is located at the northern end of the extension line of Beijing city central axis. South to the North Fourth Ring Road and north to the North Fifth Ring Road, the Olympic Park occupies a total area of 1,135hm² with an overall length of 6 km. It includes: the Northern District (the Forest Park) with an area of 680hm², the Central District with an area of 291hm², the Southern District (the original Olympic Sports Center and reserved land for sports at its south side) with an area of 114hm².

The construction projects of Olympic Park Central Zone consist of National Stadium with 80,000 seats, National Aquatic Center with 17,000 seats, National Indoor Stadium with 18,000 seats, Convention Center of 500,000m², Digital Beijing Building of 100,000m², and 510,000m² tenement buildings for athletes in the Olympic village. Olympic Park Central Zone is the core region for the 2008 Olympic Games and also the largest concentration area of stadiums in Olympic history.

The planning of the Central Zone avoids the assumption of building important projects at the end of central axis. The main stadiums are separately built at two sides of the central axis, with the National Stadium located at the east side showing a perfect contour and magnificence; the National Aquatic Center, National Indoor Stadium and National Convention Center are at the west side, reaching a dynamic equilibrium with National Stadium. This kind of dynamic and static equilibrium develops diverse space layouts of the Beijing central axis and presents the historical progress of it.

The landscape lane in the Central Zone is the extension of Beijing central axis. Its west side is planted with neat array of trees (the Green Axis) while the east side with a "dragon-shaped" water system (the Water Axis) with smooth shoreline. The whole layout forms a grand open space in the Central Zone, with central axis, green axis and water axis standing out among each other. The guarding of tree array and water system moderates the serious central axis, which highlights the "People First" idea. The design is in sharp contrast with the strictly hierarchical form of central axis in old Beijing city which the imperial palace shall be placed in the middle, with Ancestral shrine on the left and the Altar of Land and Grain on the right.

The design scheme of the Olympic Park Central Zone emphasizes the wholeness of the landscape of the Central Zone, taking the connection between the south and the north central axis space and the internal functional layout of the Central Zone into consideration and dividing the inner of the Central Area along the landscape of central axis from south to north into Celebration Square, Sinking Courtyard and Leisure Garden. Some important

Olympic facilities like National Stadium, National Aquatic Center and National Indoor Stadium are located around the Celebration Square with a high flow rate of people and various kinds of frequent activities during the Olympic Games. The Celebration Square mainly focuses on offering a wide-open outdoor plaza and then demonstrates the Olympic atmosphere. The Sinking Courtyard sections, by strengthening its three-dimensional construction of city space, provides such functions as transportation, rest, entertainment, catering, shopping, and so on, for the public and visitors after the Olympics. The Leisure Garden serves as a transition zone from the Central Zone to the Olympic Forest Park. Focusing on mixed design of artificial and natural landscape, it provides the citizens with a wide-open landscape space while satisfying public needs for entertainment, recreation and other cultural activities.

The sequence space of the central axis highlights the Celebration Square between the Bird's Nest and the Water Cube. With a length of 260m from south to north and a width of 160m from east to west, the celebration square served as the main area for celebrations and major pedestrian movement during the Olympics. The square is provided with facilities combining music and fountain, which, together with LED light show as well as the Bird's Nest, Water Cube and trees around, provides a lively but relaxing civic square.

Serving as the connection between the central axis and the important stadiums at two sides, the City Square not only demonstrates exquisite architectural forms to the stadiums, but also provides an open space necessary to evacuate in emergency, thus making the separated stadiums an organic integral space around the Central Zone square.

The central axis is designed with reference to the central axis forms of the Imperial Palace, Tian An Men Square, Temple of Heaven and south middle axis. After experts' repeated demonstration, it is finalized that the form of traditional Chinese axis should be reserved, with improvement done in details. As a result, the current 11m central axis takes shape. Considering the project characteristics, design is also made for facilities like washroom, direct drinking water, information access point, bicycle parking, signage system and other service facilities, which give back the functions of the Central Zone landscape as urban public space.

Located in the middle of the Central Zone and between the central axis and water system, the Sinking Courtyard is built as an open space and outdoor environment for the 200,000m^2 underground commercial development, subway entrance and high-flow pedestrian movement. As a result of it, the whole landscape space in the Central Zone will become three-dimensional to provide people an ideal place for traveling, shopping, leisure and meeting friends and also a new "city living room". The construction of the Sinking Courtyard provides an ingenious transition between it and the ground landscape around the central axis, that is, from ground to underground, from magnificent space to courtyard and from large square and building to Chinese-style garden.

The Sinking Courtyard is divided into two major parts by Datun Road, an east-west urban road and is connected by the municipal underpass. To offer convenience for the pedestrian movement in east-west direction at two sides of the Sinking Courtyard, three walk covered bridges are built above the south Sinking Courtyard and two above the north one. Perpendicular to the central axis, these bridges form a transport system of the chessboard shape, like in old Beijing City. Datun Road and five covered bridges divide the Sinking Courtyard into seven similar ones which form a hierarchical progressive space under the ground from south to north. Like the vertical structure of traditional residential courtyard in Beijing, the square road network and the space structure of courtyard reflect the image of traditional Chinese architectural space, thus reducing the ultra-large urban space to large courtyard space.

The design of the whole courtyard focuses on the theme "Open Forbidden City". The red palace wall of the Forbidden City is applied in the seven courtyards through breaking, malpositioning, connecting and penetrating. As a result, the whole Courtyard is outlined and its scale is adjusted.

In the first yard, huge pieces of palace wall are placed in its center. The movable "palace door" applies large-scale LED screen, which is open to the public at ordinary times. When closed, it forms a stage background, with its opposite side down into the Courtyard steps to provide outdoor seating for the public to watch games and have a rest. The second and the sixth yards, built based on the prototype of Beijing residential courtyard, respectively represent the unique understanding of classical and history. With similar prototype, they produce totally different interpretations and artistic conceptions and show the infinite possibilities of the development of Chinese classical architectures. The third yard, by incorporating the elements of ancient musical instrument-drum wall, chime and copper flute and so on, reminds people an old Chinese saying: how happy we are, to meet friends afar! The fourth and fifth yards, a transition space with less decoration crossing Datun Road, aims to reduce pedestrian stay here. The seventh yard perfectly incorporates the ramp way and steps of the Sinking Courtyard by deriving elements from Chinese classical architectures,

fine arts, decorations and so on, with a pursuit of artistic conception. It achieves a good effect.

The incorporation of traditional elements in the design of the Sinking Courtyard brings people a strong visual shock. In detail treatment, they all present uniqueness, with abundant and profound contents. It is truly a brand-new and bold try in inheriting and developing traditional culture in Chinese architecture history.

"秀"吧 北京
Xiu Bar, Beijing
2006～2009

建设机构 / Construction Organization：中国银泰集团 / China Yintai Holdings Co., Ltd.
项目地点 / Location：北京银泰中心南裙房屋顶 / Roof garden on the South Podium of Yintai Plaza
建筑面积 / Floor Area：1300m²
设计时间 / Design：2006年6月 / June., 2006
竣工时间 / Completion：2009年4月 / Apr., 2009

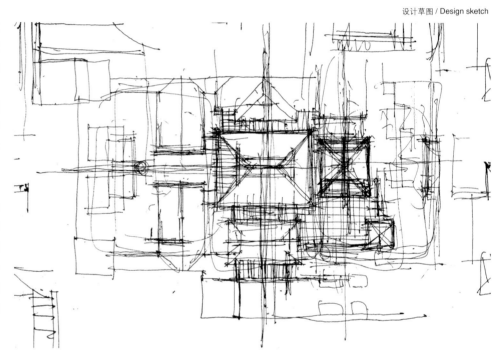

设计草图 / Design sketch

"银泰中心"位于北京市长安街与东三环路交汇处的西南方向，北侧与国贸中心对峙，是目前北京CBD地区重要的标志性项目。银泰中心包括三栋超高层建筑，其中中间一栋为高249.5米的酒店、公寓，东、西两栋同为高186米的办公建筑，三栋建筑以中间建筑为轴，对称布局。地面以上的五层裙楼为高档品牌的零售店。"秀"吧就位于三栋超高层建筑围合之下的裙楼屋顶，建筑面积仅为1300平方米。

如何在周边高层林立而又有强烈的轴线对称空间环境下展开酒吧不同功能要求和体形灵活多变的设计？我认为既然在高度现代化的建筑群中以中国传统建筑形式来表现酒吧独特的个性被视为一条可取的途径，那么寻找超大规模的现代建筑群与尺度亲切的传统建筑对话的可能性就成为设计的关键。

我注意到在传统建筑中存在着强烈的中轴线对称的空间格局，这一点在皇家建筑中表现得尤为突出。在轴线控制之下，各种体量大小、形制不同的建筑均可以串联起来，形成不断演进的序列空间。这种轴线控制下的空间关系与"银泰中心"三栋超高层建筑的布局形式有相同之处，似乎都在表达着同一的逻辑关系。于是，我将已经建成的三栋超高层塔楼强烈的轴线关系作为前提条件，进一步引申到酒吧平面中。利用居中的南北向的轴线和东西向的轴线，将酒吧各功能按照轴线空间的对位与转合进行重新布置，从而在平面和空间格局上确立与环境的对话关系。

在平面布局形成之后，我进一步强调传统建筑群中处于不同部位的建筑屋顶的从属关系，采用不同形制的屋顶，使这组建筑形成整体。这一设计方向既回应了酒吧建筑的外部形象与周围三栋高能、高层建筑的空间关系，又充分满足了酒吧建筑的品质和它灵活的使用要求。

整个建筑群由接待区、酒吧区、辅助区三个基本功能区域组成。其中酒吧以主吧（Main bar）为中心，周围环绕三个主题酒吧，包括南侧的葡萄酒吧（Wine bar）、西侧的伏特加吧（Vodka bar）和行政酒廊（Executive lounge）。接待区包括直接连接北塔的主接待厅和西侧连接写字楼的次接待厅。辅助区则由厨房、洗手间和衣帽间等组成。整个酒吧建筑与周围高层建筑之间形成了两个水景内院。

为使传统的建筑形式与酒吧功能需求相协调，避免北方传统建筑墙体多、封闭感强，突出酒吧建筑的开放与通透的特点，我们为此进行了专门的研究，并认真听取了古建筑专家傅熹年先生的意见。我们注意到在中国建筑发展过程中，宋式建筑风格是比较自由和开放的。原因是中国到了宋代手工业、商业开始发展起来，城市的繁荣与市民生活的多样性，促进了民间建筑多样性的发展；同时，"宋朝取消了里坊和夜禁制度，形成了按行业成街的情况，一些邸店、酒楼和娱乐性建筑也大量兴建起来。"宋朝建筑"比唐朝建筑更为秀丽、绚烂而富于变化"。然而，现今宋代的古建遗存极少，我们多是从宋朝绘画、雕塑等方面的文物中寻找答案。

酒吧的建筑屋顶以《营造法式》为范本，在主吧和葡萄酒吧采用了九脊顶（清 歇山顶），在雪茄室、伏特加吧、行政酒廊采用了不厦两头造（清 悬山顶），在舞台、厨房、北走廊采用了单坡加勾连搭等三种瓦屋面形式，各栋仿古建筑之间通过现代的玻璃体空间联系。建筑设计严格以宋《营造法式》对建筑柱高、屋面举折、建筑模数等方面的规定为依据。因为是按建筑平面尺寸反推算每栋建筑的"分"的尺寸，所以每栋建筑木作的模数各不相同。其中：

主吧　　1分 =12mm
葡萄酒吧　1分 =9mm

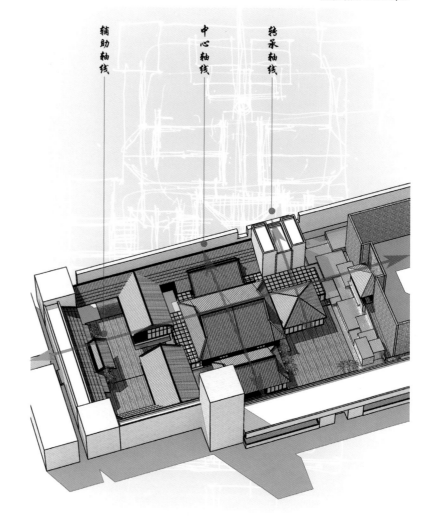

轴线分析图 / Axis analysis

分析图 / Analysis diagram

屋架 / Roof truss

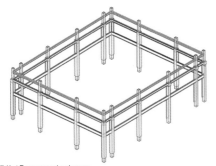

梁柱 / Beams and columns

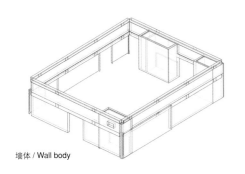

墙体 / Wall body

伏特加吧、行政酒廊、北走廊、厨房　1分 =11mm
雪茄吧　1分 =8mm
舞台及储藏室　1分 =11.5mm

因受场地和建筑物间距制约，设计缩小了主吧、葡萄酒吧歇山屋顶的檐出尺寸；北侧室外走廊单坡屋面为保证钢结构体系合理，并与相邻的伏特加吧屋面相交，没有按照举折制度计算坡度；因钢结构屋面体系的厚度远超出《营造法式》的做法要求，为保证檐口效果，采用槽钢收口来代替木连檐，并简化外檐铺作做法，通过用"单斗只替"及"替木+把绞头作"替代比较高的四铺作，降低屋架高度，弥补因钢结构厚度而升起的部分高度；因采用槽钢收檐口，为符合钢结构特性，建筑檐口角位不做生起；为满足现代功能及装修施工要求，柱的木饰面不做侧角和收分。

由于防火等级要求，建筑只能采用钢结构支撑体系，而且必须考虑钢结构材料的防火处理。所以，每一个主要杆件以及各部位杆件之间的关系都按照《营造法式》的要求，进行严格的设计。在钢结构屋架现场施工完成之后，虽然现场到处是建筑垃圾，周围是没有完工的银泰中心主体建筑，但以钢结构所建造起来的传统建筑的结构框架所表现出来的美妙之处已经跃然呈现在我的面前，让人肃然起敬。对传统建筑所蕴含的美学价值的充分认知，是通过认真依托《营造法式》进行设计才意识到的。

在钢结构施工完成之后，表面采用木材进行围合，表现木构建筑的质感。北方地区木材受气温和湿度影响变化较大，容易开裂，特别是建筑外立面部分将直接承受风雨侵袭，这是建设过程中不可回避的困难。我们就此设计了大量的技术节点进行比较，同时制作了较大比例的模型加以详细研究，最终确定了设计的细部方案。室内屋架木作均采用矩形断面，所有露明木构件、木饰面板均采用榆木板，作旧处理，见木纹，亚光清漆罩面。所有木构件、木饰面板均按古建施工规范进行防腐、防虫、防火、作旧处理。局部木构件断面尺寸太小，采用了纯木构件与钢结构翼板用螺栓连接的方式完成。

Yintai Center, which is located in the southwest of the intersection of Chang'an Avenue and East Third Ring Road in Beijing and north opposite to the International Trade Center, is now the landmark project in Beijing CBD. The Yintai Center comprises three super high-rise buildings, among which the middle one consists of hotel and apartments with a height of 249.5 meters and the other two with height of 186 meters serving as office buildings standing in the east and the west. Taking the middle one as an axis, the building group shows a symmetrical layout. Above the ground is a five-story podium building serving as a luxury brand retailer. The "Xiu" bar is located on the roof of the podium building surrounded by the three buildings, covering a building area of about 1,300 square meters.

How to unfold the bar's different functional requirements and offer a flexible design in the intense axial symmetrical space among the tall buildings? Now that it was deemed as a desirable spatial approach to represent the uniqueness of the bar in the form of traditional Chinese architecture amongst a highly modernized architectural complex, therefore the key to design, in my view, becomes the seeking of relationship possibility between the colossal modern complex and the moderately and cordially traditional ones.

I noticed that there was a strong spatial pattern of axis symmetry in traditional buildings, especially in Chinese imperial buildings. All the buildings with different scales and structures can be connected in series under the control of an axis, exhibiting a sequence space in gradual and cascading manner. This spatial relationship controlled by the axis is identical to the layout of the Yintai high-rise buildings, seemly showing the same logical relation. Thus, I took this intense axial relationship of the three high-rise buildings as the precondition and further extended it into the bar's plan. Utilizing the centered north-south and east-west axes as a guide, the functions of the bar can be rearranged in light of the spatial contraposition and transformations, thus establishing a relationship between the surroundings and layout in terms of both plan and space.

After forming the plan layout, I further emphasized the subordinate relationships between the building roofs in different locations among traditional architectural complex and adopted roofs with various shapes to unify the whole building group. This design orientation not only reflects the spatial relationship between the bar's exterior images and the three surrounding high-rise buildings but also meets its quality and flexible usage requirements.

The whole architectural complex comprises three basic functional areas, namely the reception area, the bar and the auxiliary area. The main bar is the center surrounded by three theme bars, including wine bar in the south, vodka bar and executive lounge in the west. The reception area includes the main reception hall directly connected with north tower and the secondary reception hall in the west connected with the office building. The auxiliary area consists of the kitchen, washroom and cloakroom, etc.. Moreover, two waterscape inner gardens are formed between the whole bar and the surrounding high-rise buildings.

For the purpose of coordinating traditional architectural form and functional requirements of the bar and avoiding the multi-walls and intense closure form in traditional northern buildings as well as highlighting such characters as openness and transparency of the bar, we undertook this special research and seriously took the advice of Mr. Fu Xinian, an expert on ancient architecture. We noticed that the building style in Song Dynasty was relatively free and open during the development of Chinese architectures, the reason for which lies in that the booming handicraft and commercial industries developed in Song Dynasty and the diversity of turban prosperity and citizen living styles jointly promoted the forwardness of various folk architectures; meanwhile, "Li-Fang" and "Night Ban" system were abolished in Song Dynasty, thus forming the streets by industries and also boosting the development of constructions such as inns, taverns and large entertainment buildings. The Song buildings, compared with those in Tang Dynasty, were more magnificent, gorgeous and diverse. However, only few ancient Song buildings were left, so we merely seek the answer from such historical relics as Song paintings and sculptures, etc..

西接待厅入口 / Entrance of west lobby

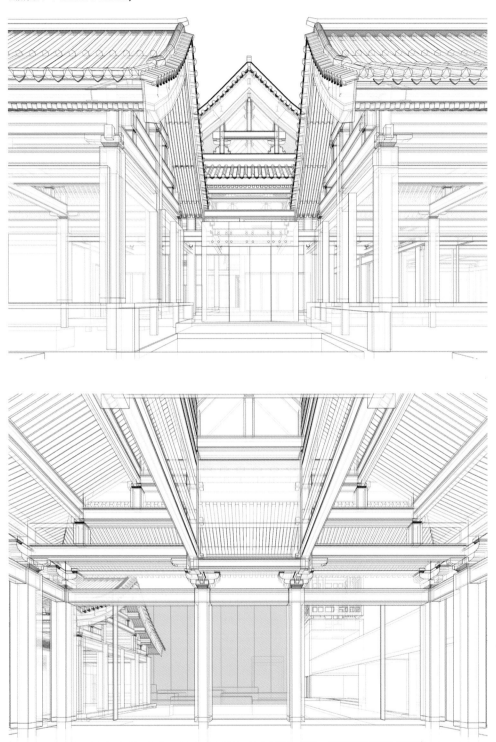

雪茄吧窗景 / View through cigar bar

工作模型 / Models

Following the model of Yingzao Fashi for the bar roof, the main bar and wine bar adopted Nine-ridge Top (referred as Xieshan Top in Qing Dynasty) while Buxia Liangtou Zao (referred as Xuanshan Top in Qing Dynasty) was used for the cigar room, vodka bar and executive lounge. The stage, kitchen and the north corridor adopted three tiling roofing in the forms of single slope and overlap. Each pseudo-classic building was linked together by a modern glass-body space. The building design is done in strict accordance with the requirements set forth in Yingzao Fashi concerning height of building column, the rise and depression of the roofing and construction module, etc.. The construction module of each building is different from each other because the basis for building plan calculation was obtained through their "fen" (a unit of length): where,

Main bar: 1 fen = 12mm
Wine bar: 1 fen = 9mm
Vodka bar, executive lounge, the north corridor, kitchen: 1 fen = 11mm
Cigar bar: 1 fen = 8mm
Stage and storage: 1fen = 11.5mm

Due to the spacing limitation between site and buildings, the eave dimension of the main bar and wine bar with leaning roof is reduced in design; furthermore, the gradient of single slope roofing at north side of outdoor corridor is not calculated per rise and depression requirements in order to ensure the rightfulness of steel structure system and overlap with roofing of adjacent vodka bar. Because the thickness of steel roofing system is beyond the method required by Yingzao Fashi, for the purpose of ensuring the cornice effect, the wooden connected eaves are substituted by trough steel binding, the auxiliary method for outside eaves is simplified, and the height of roof truss is also reduced through "simple replacement" and "alternate wood + ground for beginning" instead of the four higher auxiliaries, making good for the partial rising height due to steel structure thickness. As the adoption of channel steel eaves, the angular position of the building cornice requires no rise-up to ensure the compliance with steel structural characteristics. Meanwhile, the wood facings of columns do not need any side angle and different sizes to meet the requirements for modern function and finishing construction.

For the purpose of fire rating requirements, only steel supporting system can be applied in the buildings and fire prevention treatment of steel structure materials must be considered as well. Therefore, every piece of main rods and the relationship between rods at various locations must be strictly designed in accordance with Yingzao Fashi. Following the completion of site construction of steel roof truss, the beauty behind the steel framework of traditional buildings constructed by steel structures has already presented in front of us so vividly that it fills us with

钢结构施工 / Steel structure in construction

deep esteem notwithstanding the construction wastes everywhere on site and uncompleted main buildings of Yintai Center around. The sufficient knowledge of aesthetic values conveying in traditional buildings was achieved through the serious design relying on Yingzao Fashi.

After the completion of steel structure construction, the timber was applied for surface covering to show the texture of wooden architectures. In northern region, the timber is prone to temperature and humidity, thus easily being cracked, especially the elevation of building which is directly exposed to weather, causing unavoidable problem during construction. For the purpose of the said issue, we designed abundant technical nodes for comparison; in addition, we also made large models for further detailed research, and developed the final detailed design scheme. Indoor roof truss applied with rectangle section and all exposing wooden members and veneers were made from elm, disposed with reconditioning and applied with semi-gloss varnish top facing, the wood grain of which can be viewed. All wooden members and veneers underwent corrosion prevention, insect prevention, fire prevention and reconditioning treatment and complied with ancient architecture construction specifications. Since the partial wooden structures section was too small, it was completed by bolting with pure wooden components and steel flanges.

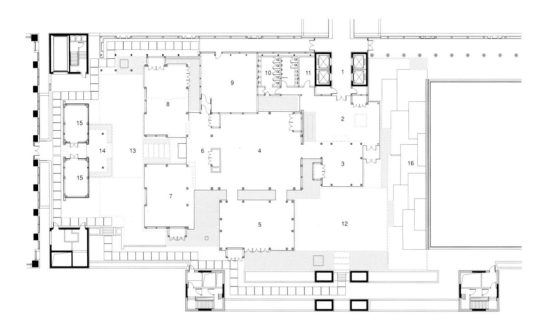

1	电梯厅	9	厨房
2	东接待厅	10	卫生间
3	雪茄房	11	存储间
4	主吧	12	东露台
5	葡萄酒吧	13	西露台
6	西接待厅	14	舞台
7	行政酒廊	15	储藏室
8	伏特加吧	16	水景

总平面 / Master plan

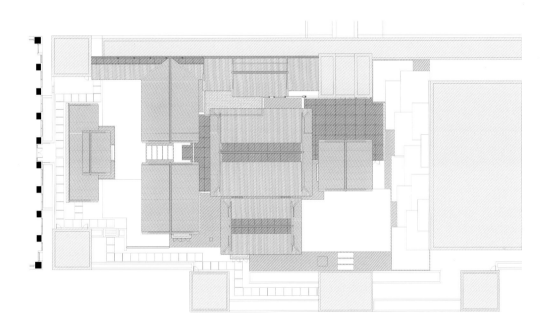

屋顶平面 / Roof truss

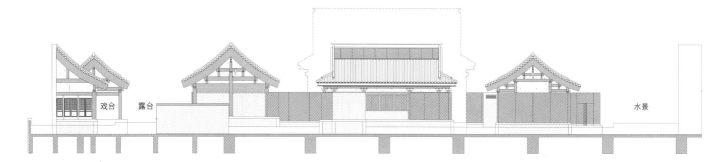

E1 立面 / E1 Elevation

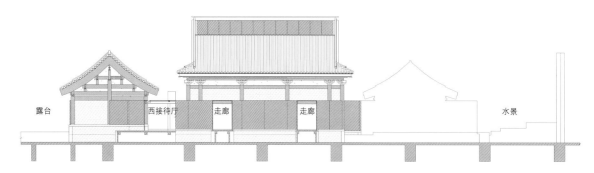

E2 立面 / E2 Elevation

E3 立面 / E3 Elevation

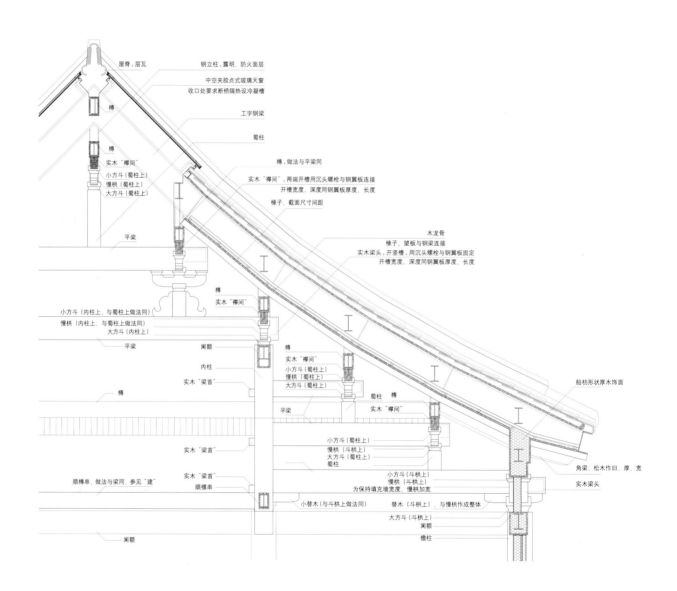

主吧木作详图(横剖面) / Details of wooden wares of Main Bar (cross section)

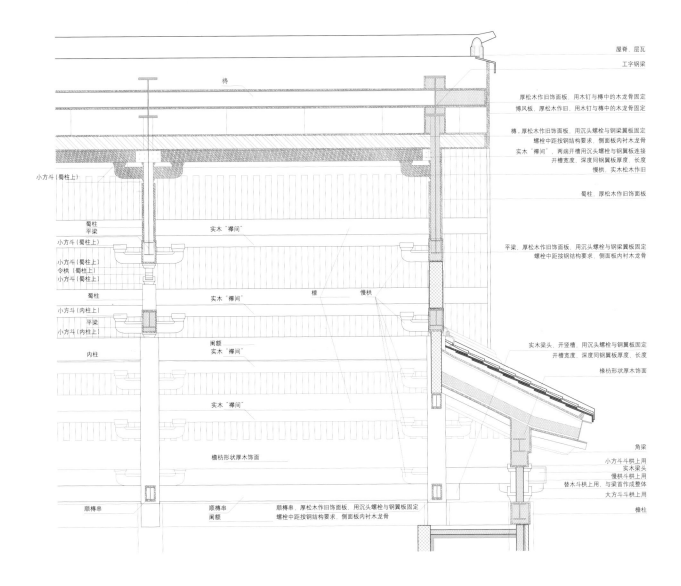

主吧木作详图（纵剖面）/ Details of wooden wares of Main Bar (longitudinal section)

哈德门饭店　北京

The Reconstruction of Hademen Hotel, Beijing
2006

建设机构 / Construction Organization：国瑞集团 / Glory Group
项目地点 / Location：北京崇文门 / Chongwenmen, Beijing
建筑面积 / Floor Area：160602m²
建筑高度 / Builging Height：60m
设计时间 / Design：2006年6月 / June, 2006

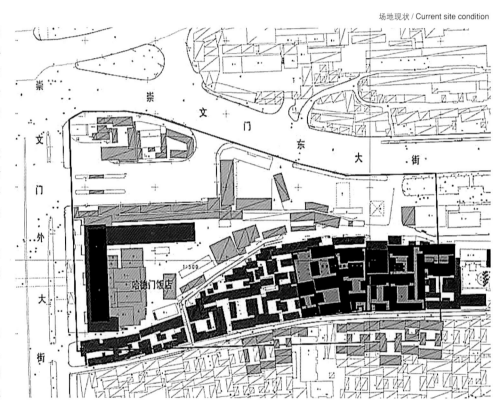

场地现状 / Current site condition

哈德门就是现在的崇文门，哈德门饭店在崇文门的东南方向。饭店是1974年建造的，只有34年的历史，属于当代建筑，谈不上文物保护。而哈德门地名古已有之，当时只不过使用它进行命名，但并不包括哈德门的历史。过去哈德门饭店有几大优势：一是离市中心近，走着就能到天安门、王府井；二是离北京火车站近；三是著名的便宜坊烤鸭店就在饭店一层，请客方便。后来随着新世界中心、国瑞城的建立，逐渐使哈德门饭店开始显得有些土气而不合时宜。于是崇文区政府决定将哈德门饭店连同周围用地一起进行土地开发，拆除现在的哈德门饭店，建设高标准的新哈德门饭店。土地及开发引进房地产商完成，并有可能成为项目的最终建设方。

这是一块很不规则的建设用地，北侧的崇文门东大街和西侧的崇文门外大街决定了用地的形状。用地内除原来的哈德门饭店外，还有一些不同时期的建筑，即将被拆除。通过对基地现状的研究，我注意到在用地的东南角有保存基本完好的6个院落，文物部门要求将这六个院子进行异地重建。于是我就想为什么不能在原场地上重新建设这6个院子呢？答案很简单，因为该场地上将要建设新的哈德门饭店，由于占地面积较大，不可能为这6个院子的保留空出相当的场地。但是，我又提出了新的建议，为什么不能在新的建筑中保留这些院子呢？这是一种新的解决途径，文物部门从来没有研究过这种方法，原因很简单，没有开发商愿意保留这些并不很有价值的老房子，规划管理部门也没有将原本是文物部门管理的保护型房屋纳入管理的范围，而开发商宁愿花更多的钱，在新建筑中尝试建设新的传统建筑，也不会想到平时总是向他们提出严格、甚至是过分要求的文物部门会同意将保留的房屋允许留在他们的场地之中。这就是我们社会的现状，相互割裂的城市空间资源的管理与建设格局，使城市空间的品质无法提升。因此，这6个院子不应当成为展开设计的障碍，最佳的解决方案是找到原有建筑与新建筑之间对话与共存的设计途径。

场地原状 / Original site

场地北侧的明城墙遗址 / Relics of the city wall of Ming Dynasty to the north of the site

城市建筑的体量变化 / Volume variation of city buildings

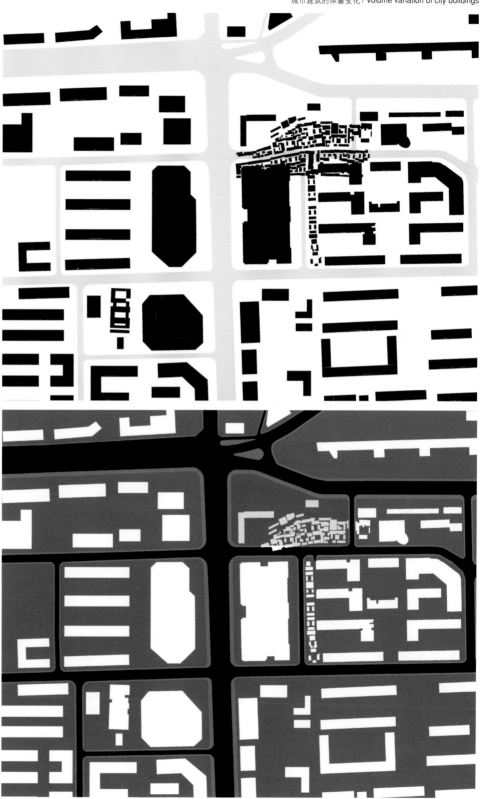

北京旧城之中，在大体量综合体建筑不断出现的同时，也伴随着传统低层院落的迅速消失。此消彼长的过程中，两者之间存在的巨大的尺度落差，既体现在建筑实体，也体现在发生于其中的生活状态和公共行为方式上。崇文门商业区逐步向商业、居住、公共建筑等多种机能叠加的复合城市区域转化，积聚着由多样化、选择性的城市生活蕴含的巨大能量。

在城市急速更新的过程中，新与旧、大与小两者之间平稳过渡的缺省，导致了我们对城市文化连续阅读的障碍。因此，将不同的空间形态包并置起来，突出了两者之间互立、互含的关系，强化时空与体量变化的冲突，从而形成异质空间之间特有的张力，使设计从二元的关系中延伸出来，具有阶段性的文化意义。

我希望以包容的设计理念，收藏传统形态建筑的手法，重新构建新的场所精神，超越对四合院简单的保留和保护，打造多种复合功能为一体的酒店综合体建筑。

在一处具有历史意义和文物保护的环境内，现代城市建筑均以庞大的体量来标定已经不断突破的尺度，这是不可避免的。但除此之外，新建筑还有其他什么意义值得建筑对其精雕细刻，一定要体现出新建筑存在的价值吗？我认为没有必要，新建筑既然

城市空间的尺度变化 / Size variation of city space

保护院落原貌（图1-6）/ Original scene of preserved courtyards（Figure 1-6）

必须具备这样庞大的体量，那就让它承担起勾勒城市轮廓的作用，以最简单的体形、轮廓和细部处理作为设计的背景。

于是，新建筑的设计首先是对各功能的需求进行详细的设计，找到符合功能要求的体量关系。整栋建筑沿东西方向排开，并被挤压成局部围合的状态，恰好与各功能布局和需要相一致。在不同的部位结合结构的可能性，尝试着留出不同跨度的巨构空间，为下一步完成新、旧建筑的"混合"创造条件。

在对6个院子进行认真清理和研究的基础上，确定了将要保护的范围、轮廓和房屋的格局。并将每个院子按照新建筑的进深和跨度的可能性，将这些院子进行旋转，并最终调整成南北和东西两个主要方向，而原来每个院子的微小的偏离角度被保留下来，体现现代建筑与传统建筑在空间方位上的拼贴关系。接下来将每个房屋植入事先已经准备好的巨大的现代建筑空间之中，并赋予新的功能。

巨大的跨度空间在不同方向成为城市空间的一个过渡，而传统院落与新建筑的并置则强调了城市空间的文化与历史属性。新建筑不再是孤立的，它通过城市空间的塑造将承载城市历史的保护院落收藏在其中，将建筑与城市紧密联系在一起。

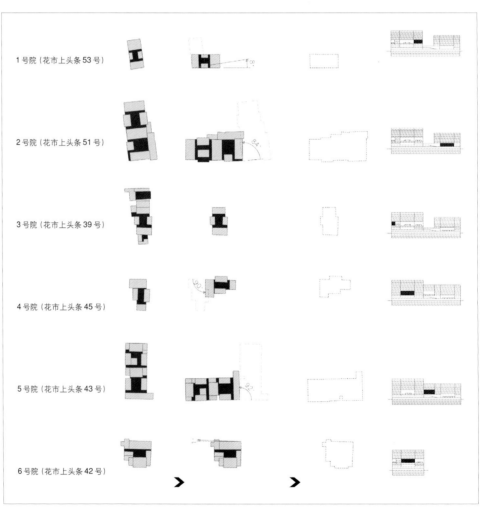

院落重置 / Courtyard reset

公共空间及路径设计 / Conception generation of public space

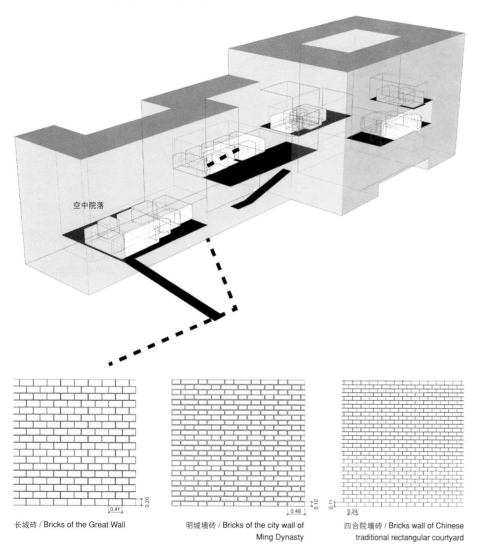

长城砖 / Bricks of the Great Wall

明城墙砖 / Bricks of the city wall of Ming Dynasty

四合院墙砖 / Bricks wall of Chinese traditional rectangular courtyard

　　复合区域的城市活力和机能的实现，依赖于综合体建筑本身功能的多样化和丰富性，以及与此相关的内外公共空间区域的连续性和可达性。设计在完成基本构成之后，以开放的姿态设计城市新的公共文化休闲场所。在有限的基地范围内，在沿崇文门东大街的广场巧妙地创造地形，形成和缓的下沉台地广场，广场地面的两座桥梁将人流分别引向一层展厅、会议中心和二层空中院落，与酒店前广场和北面明城墙遗址公园一起，构成连续的城市公共空间。

　　建筑外立面采用了中国传统建筑围护材料——砖墙的颜色和砌筑图案，以不反光的石材墙体构成完整的立面效果，突出新建筑的体量感。在空中院落两侧外墙处采用了与四合院的瓦坡顶质感相类似金属波纹板作为外墙材料，与大面积石材墙面形成对比，强调了建筑的时代特色。

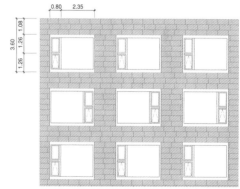

立面设计 / Facade design

Hademen is the present Chongwenmen, and Hademen Hotel lies to the southeast. The hotel constructed in 1974 and having a history of only 34 years is part of contemporary architecture and is in fact far from a cultural relic to be preserved. However, the name of Hademen has been in existence since ancient times. It was used only as a place name then and did not have a history concerning Hademen in particular. In the past, Hademen Hotel had several unique advantages: first, it was close to the downtown area, within walking distance from Tian'anmen and the commercial district of Wangfujing; second, it was close to the Beijing Railway Station; third, the famous Bianyifang Roast Duck Restaurant was located on the first floor of the hotel, which made it convenient to entertain guests. Then with the establishment of New World Center and Glory City, Hademen Hotel gradually became boring and out of date. Therefore, the government of Chongwen District decided to redevelop Hademen Hotel together with its surrounding land by removing the old Hademen Hotel and building a new one with higher standard. Real estate developers have been brought in and are very likely to become the final constructors of the project.

This is a rather irregular construction site. The Chongwenmen East Street in the north and Chongwenmen Outer Street in the west determine the shape of the site. Besides the original Hademen Hotel, there are architectures of different times, which are to be demolished soon. By investigating the current site, I noticed that in the southeast of the land, there are 6 basically well-preserved courtyards, which are to be reconstructed on another site as required by the department of cultural relics. However, an idea came to me that why the reconstruction can't be conducted on the same site. The answer is clear and cut. It is because that the new Hademen Hotel taking up a large area is to be reconstructed on the site and enough land can not be reserved for the 6 yards. However, I put forward a new suggestion that why these yards can't be preserved among new architectures, which is a new solution that the culture relics department never considers. The reason is simple: no developers are willing to preserve such old houses with little value and the planning management departments do not regard the protection of houses concerning the departments of cultural relics as their responsibility. Furthermore, developers would rather spend more money in trying to build new traditional architectures among new architectures, but would not think that the departments of cultural relics posing strict and even demanding requirements on them would agree preserved houses on their properties. This is our social situation that the divided resource management of city space and construction pattern makes it impossible to update the quality of city space. Therefore, the 6 yards should not become obstacles for design. The best solution is to find the creative approach that enables dialogue and coexistence between original buildings and new ones.

In the old city of Beijing, large-dimension building complex constantly emerges accompanied by quick disappearance of traditional low-rise courtyards at the same time. During the process, great differences between the two are reflected not only in construction entities but also in living conditions and public behaviors. The commercial district of Chongwenmen is gradually transforming into composite urban areas with various functions such as commerce, residence and public buildings, and accumulates huge energy contained in diverse and selective city life.

During the rapid process of urban renewal, the lack of smooth transition between old and new and large and small led to obstacles to our continuous reading of city culture. Therefore, the juxtaposition of different space forms and highlights the mutually independent and mutually included relationship between the two and intensifies the conflict between space & time and dimension. As a result, peculiar tension is formed among heterogeneous spaces so that the design can extend out of dualism and bear a periodic cultural significance.

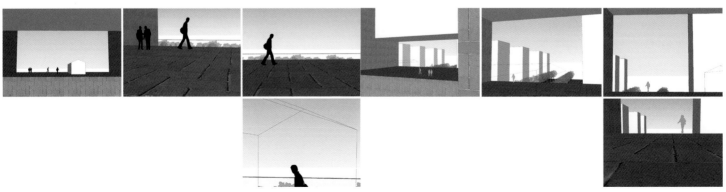

庭院视线分析 / Analysis of courtyard sight　　　　庭院视线分析 / Analysis of courtyard sight

I hoped to incorporate the traditional architectural forms and reconstruct a new sprit of the site by an inclusive design concept a concept far beyond the simple preservation and protection of the courtyard and finally build a hotel complex integrating various functions.

In a historic and relics protection environment where it is inevitable that modern urban architectures all take the large dimension to demarcate the increasing growing scale. However, besides this, whether there are other aspects concerning new buildings that worth being treated with care by architects so that the existence value of new architectures can be reflected. On my part, it is totally unnecessary. If the new buildings must be huge in dimension, let them take the role of outlining the contour of city with the simplest shape and contour and detail treatment as design background.

First of all, the new building design carries out a detailed design according to demand for each function, so that the relationship of dimension that meets functional requirements can be found. The whole building is constructed from east to west and pressed to be a partial enclosure exactly in accordance with layout and requirement of each function. Study the possibility of integrating structure in different positions and try to make huge space with different spans, so as to create conditions for the next combination between the old buildings and the new ones.

Based on the careful cleaning and study of 6 yards, the scope to be protected, contour and house patterns are determined. In accordance with the possibility of the depth and scope of new buildings, these yards are rotated and finally adjusted into north-south and east-west directions while the original small deviation angle of each courtyard is retained to embody the collage relationship in spatial position between modern architecture and traditional architecture. Next, each house is implanted into the well prepared large modern building space and endowed with new functions.

Large span space becomes a transition of city space in different directions while the juxtaposition of traditional courtyard with new building emphasizes the cultural and historical attribute of city space. The new building is no longer isolated but incorporates the protected yards bearing city history through shaping city space and links building and city together closely.

The realization of urban vitality and function of composite regions depends on the functional diversification and richness of building complex and the continuity and accessibility of related internal and external public space areas. After completing the basic composition, the design devises new city public culture and leisure places with an open posture. Within the limited base, the square along the Chongwenmen East Street ingeniously creates terrain to form gentle sinking platform square. Two bridges on the square direct pedestrian flow into the exhibition hall conference center on the first floor and air courtyard on the second floor respectively. Together with the square in front of the hotel and the Ming City Wall Ruins Park in the north, they form a continuous city public space.

The exterior of the building uses traditional Chinese building maintenance materials, with brick color and masonry pattern created by non-reflective stone wall to form complete elevation effect and highlight the dimension sense of the new building. The outer wall located on both sides of air courtyard adopts metal corrugated plate similar to tiles used in the courtyard as its materials, forming a sharp contrast with large stone wall and emphasizing architectural characteristics of the times.

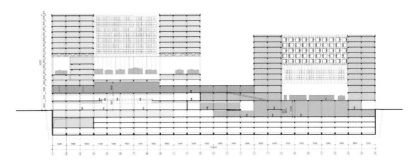

剖面 / Section

"旬"会所 北京
Xun Club, Beijing
2007～2010

建设机构 / Construction Organization：个人 / Person
项目地点 / Location：北京百子湾路 21 号院 / No. 21, Baiziwan Road, Beijing
建筑面积 / Floor Area：1664m²
设计时间 / Design：2007 年 8 月 / Aug., 2007
竣工时间 / Completion：2010 年 8 月 / Aug., 2010

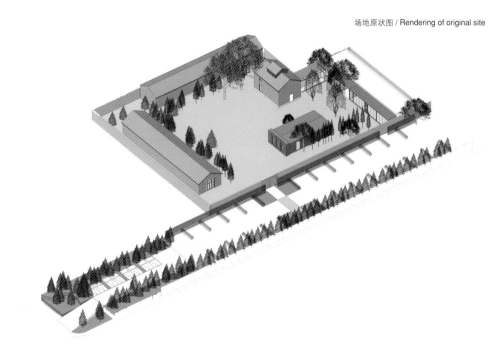

场地原状图 / Rendering of original site

这是一处旧有的车站维修用房，形成年代不详，只因为产权归属于北京铁路局，所以才没有被拆迁而在北京高速发展的城市更新过程中保留了下来。原有建筑破败不堪、落满灰尘，是新的主人将这个院子租用过来，并打扫、整理一新。以至于有铁路领导再来视察时，对这样的变化不禁表现出十分惊愕的样子。

院子里由南至北大体分为三部分，南侧是一栋两坡顶木屋架平房；位于中部的其中一栋是平房，另一栋是一个带阁楼的两坡顶小楼，顶部有一个采光窗；北侧是一栋改造过的单坡顶平房；西侧这栋是一路南北走向的西房，恰好将这三部分建筑联系在一起。院子中最有特点的是几十年来逐渐形成的绿化环境，以乔木为主，主要是松树和梧桐树，还有一些观赏树木。院落北侧有三棵较大的挂牌保护的古树，树木的位置布局恰到好处、姿态也非常优美，与建筑相得益彰，这不禁让初到这里的人们感到新奇。虽然这个院子紧贴东四环，但在大树的庇护下，却显得非常幽静。

设计是从功能的梳理和确认开始的。会所建筑设计的难点在于要求建筑师具备功能的理解和运营的经验，才能根据建筑的不同特点规划出适宜的使用功能布局，也以此成就会所的个性特征，获得市场的认可。根据现场建筑的特点，南部的坡屋顶房屋安排展览的功能，其颇具历史感的人字形屋架增添了几分沧桑；北部的单坡屋顶的房屋由于必须保留横隔墙，不能整体使用，最适合作为 VIP 房间；西侧建筑可以安排厨房和其他辅助功能房间；而中部的建筑由于面积和空间的限制，很难直接派上用场。

中部原有建筑随意散落的状态既不适应会所的功能需要，也无法阻止会所的人流动线，我提出了添加一个主要空间的想法，它将统领各个功能空间，起到汇集人群、联络各处的作用。我试图通过一个正方平面格局的新建筑镶嵌到院子中部、二层坡屋顶小楼的南侧，东面与小平顶相连通，形成有新、旧三栋房子组成的中心，这里将是会所的接待空间、主吧空间和酒窖、雪茄吧，在平面功能上与南、北两部分原有建筑形成比例适宜的会所功能。主吧的建筑是整个院子中唯一一栋新建筑，为了突出这栋

卫星视图 / Satellite view

区位图 / Site location

建筑的地位和新材料与原有建筑的不同，同时也考虑院子中心观赏角度的需要，我将新建的主吧特意向东旋转了一个角度，造成了在外观上与旧有建筑空间格局的冲突。新建筑在外观上采用了最低廉的瓦楞铁皮的材料，在阳光的照耀下泛着微弱的银色的光，特有的质感给院子带来了生气。建筑的外观体形也是独特的、甚至是不稳定的、顽皮的样子。逐渐升起的屋顶旋转着指向西侧两颗巨大的梧桐树方向，如同欲跑到树荫下躲藏的孩童。的确，屋顶的一部分可采用玻璃屋面，坐在主吧里享受时光的客人可以透过玻璃屋顶看到室外天空的变化，这面玻璃屋顶也正好位于西侧梧桐树的阴影之下，为建筑室内遮蔽阳光。不幸的是，在后来施工的过程中业主发现前面的一棵梧桐树由于树龄太大，树干上面分叉的地方已经要坏死，如果不锯掉的话，有可能哪一天掉下来，砸坏主吧的玻璃屋顶或当中的客人。唉！原指望用树影来遮阳的，可没想到聪明反被聪明误，梧桐树的上部被锯下来后，玻璃屋顶的遮阳还得重新考虑。幸好当时拍了照片，就算是一个美好瞬间的永久纪念罢了。

由于会所的主要出入口只能设置在东侧，且开口位于南侧和中部建筑之间，如果客人直接进入院落，院落中的主要空间和建筑将尽收眼底，缺乏神秘感和吸引力。所以，我借用了北京四合院中，进

场地原状 / Original site

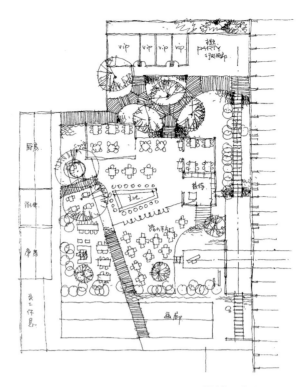

设计草图 / Design sketch

入垂花门之后，用"仪门"遮挡，引导客人从两侧的"抄手游廊"进入庭院各个房间的手法，在主入口前设计了一道钢框玻璃门，玻璃采用竖线条磨砂处理，两层玻璃错位安装，形成正面阻隔视线的作用。在大门两侧的甬道外侧采用金属帘作为轻隔断，部分阻隔客人视线，维持神秘的感受。由此，将客人引导到北侧的接待空间或南侧的展览空间。

在接待空间和展览空间的入口处，我设计了"喇叭口"状的半室外通道与南北方向的甬道相接。一方面考虑从城市到此的客人在进入会所建筑室内之前从空间和心理上有一个过渡，另一方面也是由于相关的建筑尺度较小，与门廊、甬道在高度和开门的方面无法正常连接的原因。

设计中我始终考虑着原有的很有价值的树木如何使用的问题，这是与建筑设计相对应的另一半工作。我们在充分测量的基础上，小心翼翼地为每棵树设计最好的建筑景观，将树干、树枝、甚至树叶和树影都作为设计的要素，加以利用。所有联系各功能房间的通廊都与景观密切结合，尽可能穿行其间，让客人足不出门就可感受到环境的存在。建筑与绿化环境的和谐共生，充分表现出会所建筑高情感的建筑特性。使新、旧建筑融入到景观环境之中，也使客人自然地融入到这样的氛围之中。

会所的主入口与建筑的空间关系确定之后，我试图在围墙上建设高度将近10米的隔声板，将东四环路上的机动车噪声全部或部分地反射回去。但此项努力没有成功，原因是建造的难度和成本，后来不得不放弃改为在原有3米高的围墙上加装2米的玻璃，尽可能地作一些隔离。入口处结合"仪门"的考虑，设计了面向庭院高大的构筑物，在形象上取传统建筑牌楼的意向，采用了一排不锈钢管作为"瓦屋顶"的象征，成为南部庭院中主要的视觉焦点。同时高悬的钢管在微风吹动下随意摆动，相互碰撞，发出清脆和谐的声响，如同风铃一般，也可平抑一些外部的噪声干扰。夜幕降临，来自西侧的一束灯光会照在"牌楼之上"若隐若现，似明月初上高楼，为庭院中活动的客人创造另一番诗意的遐想。

主入口的门廊从"仪门"直接延伸到东侧前院的围墙处，以一处方正的水池结束，池水中的磐石之上涓涌的清水，映着阳光，表达着会所高雅的格调。门廊上随风飘曳的装饰织物在夜晚灯光的照耀下，发出变幻的光芒，欢迎来自城市各个角落的客人。主入口外面安放了一对上马石狮雕，"仪门"的外侧安放了一对抱鼓石狮雕，与钢材、玻璃等现代建筑材料形成鲜明的对比，提示时空变换的印象，同时也是增强商业活动的氛围。

室内设计尽可能采用纯白的涂料墙面，为今后逐渐增设的艺术品收藏创造展示的可能性，也将原有建筑的木结构屋顶和核心建筑的钢结构屋顶通过白墙的衬托进一步展现出来。大面积地面采用了自流平水泥地面，颜色较深，与白墙面形成反差，适应今后商业运营和各类艺术活动的需要。在一些重点区域，如主吧的酒吧台、展览空间的中间部位以及接待空间的北侧墙面处均采用了我本人多年前收集的一些中国传统建筑的藏品加以装饰，使现代纯净的空间形态与极富装饰性的中国传统建筑构架形成了视觉和文化思想层面的反差，形成会所的鲜明特色。

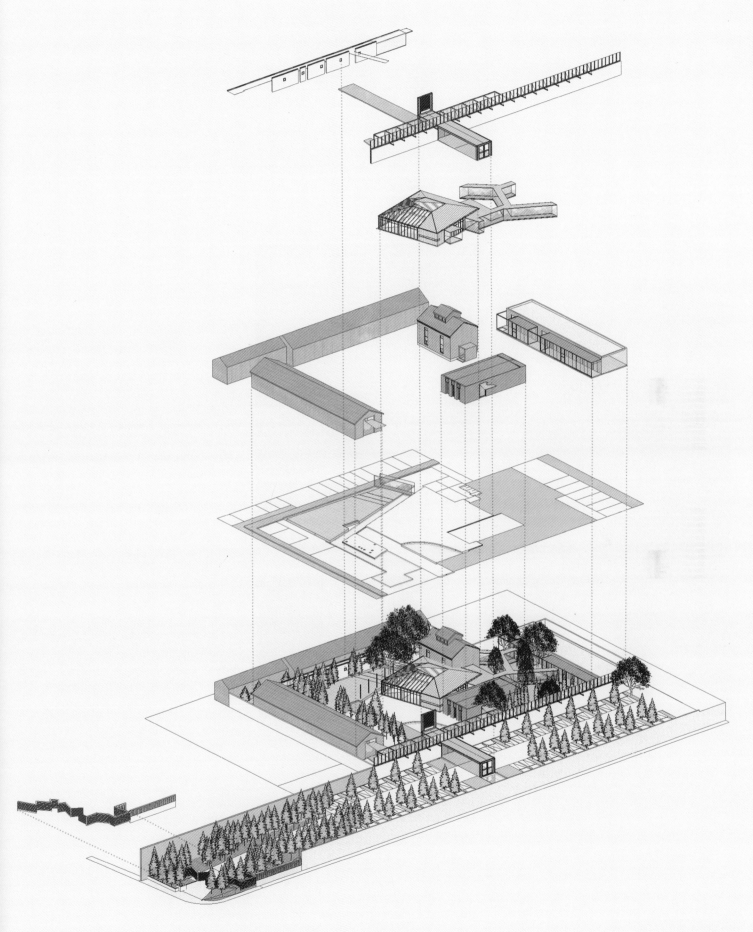

功能分析图 / Analysis diagram

建设过程 / Construction progress

This is an old station maintenance building with no clear formation age. It has not been removed while preserved in the booming process of urban renewal, for the only reason that the property right belongs to Beijing Railway Administration. The original building was dilapidated and dusty, and the new master rented and cleaned it, therefore changes there surprised leaders from the railway station when they visited there again.

The yard mainly consists of three parts from south to north. At the south side is a single story building of wood roof truss with double pitched roof; a single story building and a small building of double pitched roof with a cockloft stand in the center and a lighting window on the top; a single story building of reconstructed single pitched roof is located at the north side. The three parts of buildings were connected by a western building arranged from south to north at the west side. The yard is characterized by tens-year old trees with arbors, which mainly are pine and phoenix trees as the major, and some ornamental trees. Three giant ancient trees officially protected are located at the north side of the yard. The perfect location and layout as well as graceful posture are complement with buildings, by which, people who firstly comes here are always fascinated. Shielded by giant trees, the yard is very quiet and secluded although nearly closed to East 4th Ring Road.

The design began with a functional arrangement. The difficulty in designing club building is that only when architects understand the functions and have experience in operation can appropriate functional layout be worked out according to diverse features of the building and personality characteristics of the club be showed and finally accepted by the market. According to characteristics of building on site, the south house with pitched roof is used as exhibition room, and its herringbone roof truss with a great sense of history adds a feeling of vicissitudes. For the north house with pent roof, because it has to preserve its cross wall, it can not be used as a whole, but suitable for a VIP room. Buildings in the west can be arranged as kitchen and for other auxiliary functions. Buildings in the center can not be directly used due to limited area and space.

Scattered original buildings in the center are not appropriate for functional needs of the club or stopping people's move, so I proposed adding a main space which will play a significant role in joining all spaces together to gather people and connect with other parts. I tried to add a new square construction in the middle of the yard and the south side of two storey small building of pitched roof, with its east side connected with small flat-roofed house to form a complex of three old or new buildings. It will be arranged for reception, main bar space, wine cellar and cigar bar of the club and creating functional spaces with proper proportion with original buildings at south and north sides in terms of plane functions. The main bar is the only new building in whole yard. In order to highlight different status and new materials with original ones, and meanwhile take into consideration the view from the center of the yard, I deliberately rotate the newly-built main bar by a degree to east, giving rise to a visual conflict against original buildings in spatial pattern. Covered by the cheapest corrugated iron sheet, the new building sparkles with silver light in the sun and the special texture makes yard vibrant. The appearance of building is also unique, and even unstable and mischievous. Gradually rising roof points at the two giant phoenix trees in the west, like a child who is going to run and hide under the shade. It is sure that part of the roof can be constructed with glass roof covering. Guests sitting in main bar and enjoying themselves can catch the sight of changes in the open air through glass roof. And this glass roof is just under the shade of phoenix tree in the west and keeps out sunshine. Unfortunately, the owner found in construction that necrosis occurs at a trunk bifurcation on a phoenix tree due to aging. If the trunk is not sawed off, it would fall down and drop on glass roof or guests there. I was hopping to use it to keep sunshine, however, I have outsmarted myself that when the upper of phoenix tree was sawed off, I had to reconsider the sunshade of the roof. Fortunately a picture, a sort of lasting memorial for the beautiful moment has been taken.

Because main entrance of the club can only be in the east and the opening is suited between buildings at the south side and in the center, guests entering the yard will have a panoramic view of the main space and buildings there, without no mystique and attraction left. However, in Beijing courtyard, guests entering festoon gate will be kept out by "Yi Men" first, and then led into

rooms in the yard by "Chaoshouyoulang" at both sides. A steel-frame glass door is located at the main entrance. The door which is constructed by two dislocated layers of matted glass processed in vertical moulding blocks out sight from the front. Metal frame is used at external side of the course way at both sides of the gate as partition to block part sight of guests and maintain the mystique. From there, guests are led into reception room in the north or exhibition space in the south. At the entrance of reception and exhibition space, I designed trumpet-like half outdoor access, which is connected to the course way in south-north direction. The reason lie in that it gives guests a transition in space and psychology before they enter the property from outside, and that relevant building is small in size and can not be normally connected to the porch and course way in height and door opening.

During design, I always considered how to make use of the original valuable trees, which is the other half work corresponding to architectural design. On the basis of adequate measurement, we cautiously designed the best landscape for each tree and incorporated trunk, branch and even leaves and shadow as design elements. All vestibules connected with various functional rooms are closely united with landscapes and pass through them if possible, with an intention to provide the guests an experience of environment when staying indoors. Harmony and coexistence of architecture and greening environment fully represent the high touch as architectural feature of the club building. Moreover, new and old buildings are incorporated into landscape to assimilate guests into this atmosphere naturally.

When the spatial relationship between main entrance and building of the club is determined, I tried to construct a soundproof panel of about 10m in height on the perimeter wall, with an aim to fully or partly reflect the noise from motor vehicle at East 4th Ring Road back. However, it did not work due to the difficulty and cost of construction. And I had to give up the original design and constructed a 2-meter-high glass wall on the original 3-meter-high wall, and at the same time, added some insulated panels where possible. Considering "Yi Men" at entrance, a tall structure facing the courtyard is constructed there. On image, it selects decorated archway of traditional building and applies a row of stainless steel tube as symbol of "tile roof" which becomes a key visual focus in south courtyard. While the steel tubes hanging high sway in the wind, they crash into each other and make a nice ringing sound. They like wind chimes and also can moderate some disturbance from external noise. When night falls, a beam of light from the west sheds above the decorated archway and makes it indistinct. It looks like moon climbing on building and creates a poetical reverie for guests in the yard.

Porch at the main entrance directly extends from "Yi Men" to the perimeter wall in east front yard, and ends at a square water pond, in which, the clean water is flowing over stone and reflecting sunshine, showing the elegant style of the club. Ornament fabric swaying with wind on the porch shines with baffling radiance under light in the evening and welcomes guests from various parts of the city. A pair of mounting carved stone lions is located outside the main entrance and a pair of carved stone lions who are holding drums is located outboard of "Yi Men". They stand in sharp contrast to steel, glass and other modern building materials, showing a strong sense of space and time transition and enhancing the atmosphere of commercial activities.

Pure white painting is applied for interior design, so that it is possible to exhibit art collections which are gradually increased in the future and also to further set off the steel-frame roof of wooden-roof main building for original building against white wall. Self-leveling cement in deep color is applied at large scale, creating a stark contrast with white walls to meet any needs of business operation and artistic activities. Some important locations, such as counter of the main bar, the center of exhibition space and north wall at reception area, are decorated with my personal collections of traditional Chinese architecture, making sharp contrast in visual and cultural aspects between modern pure spatial pattern and traditional Chinese architecture and creating a club with distinctive features.

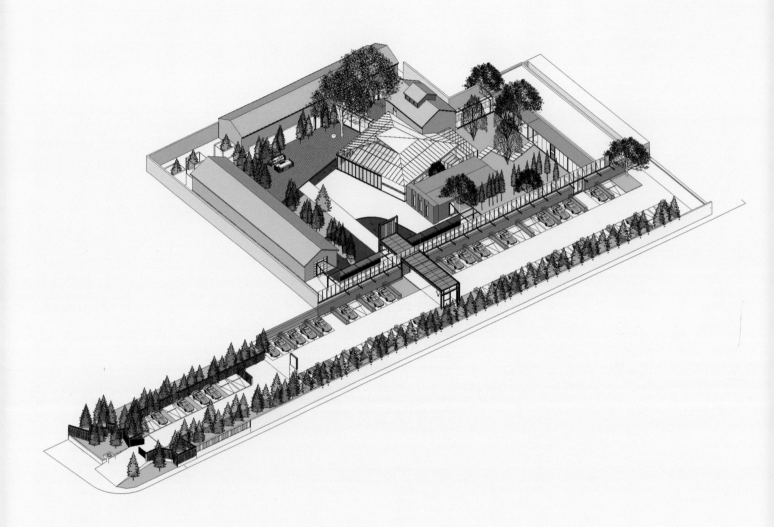

1	主吧
2	雅间
3	藏酒
4	舞台
5	接待
6	存衣
7	结算
8	办公室
9	洗碗间
10	副食操作间
11	备餐区
12	二次更衣
13	冷菜间
14	加工间
15	冷冻库
16	冷藏库
17	配电室
18	锅炉房
19	员工休息
20	工作室
21	后台
22	展厅兼媒体发布
23	画廊
24	通道
25	宿舍
26	VIP
27	雪茄吧

总平面 / Master plan

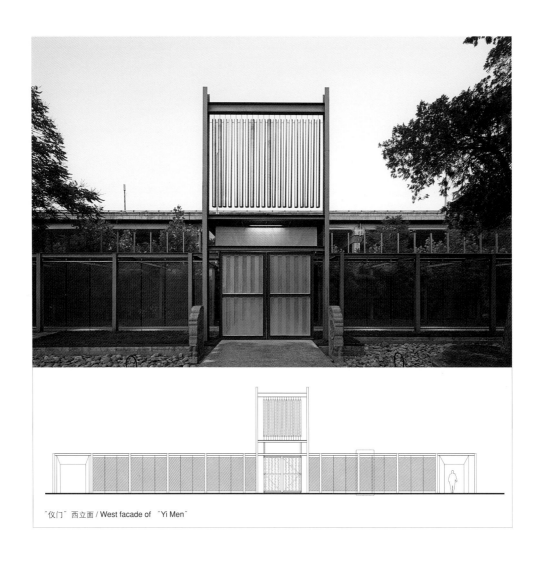

"仪门"西立面 / West facade of "Yi Men"

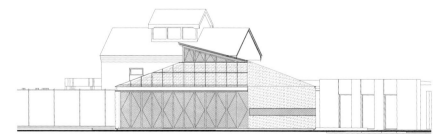

主吧南立面 / South facade of main bar

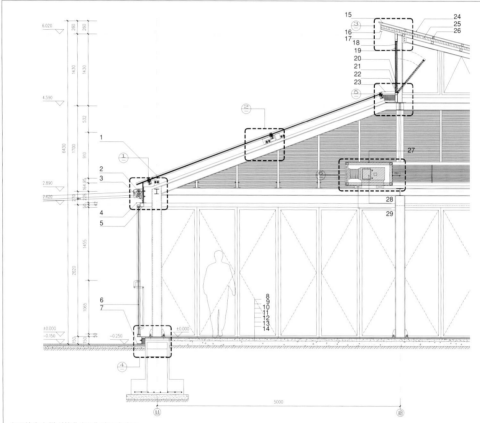

主吧墙身大样 / Wall detail of main bar

1 双层 Low-E 玻璃；2 solidux 凉棚；3 双层 Low-E 玻璃；4 坡水板，不锈钢；5 方钢 80mm×80mm，5 厚，表面深灰色氟碳喷涂；6 双层 Low-E 玻璃；7 通长槽钢 10050，7mm 厚，表面深灰色氟碳喷漆；8 1-2 厚自流平环氧面漆涂层；9 环氧漆底涂一道；10 20 厚 1：2.5 水泥砂浆找平压实赶光；11 4 厚水泥一道（内掺建筑胶）；12 50 厚 C10 混凝土；13 100 厚 3：7 灰土；14 素土夯实；15 镀锌铁皮装饰屋面，原色；16 滴水；17 金属装饰板；18 双层 low-E 玻璃；19 幕墙竖向龙骨，表面深灰色氟碳喷涂；20 内旋电动开启窗，窗框表面深灰色氟碳喷涂；21 幕墙竖向龙骨，表面深灰色氟碳喷涂；22 双层 low-E 玻璃；23 金属盖板，厚度 2mm；24 镀锌铁皮装饰屋面，原色；25 100 厚拼接夹芯屋面板；26 龙骨；27 空调装饰罩，3mm 钢板，外喷深色装饰漆；28 空调风机；29 减震胶垫

接待厅南立面图（改造前）/
South facade of reception hall (before renovation)

接待厅东立面图（改造前）/
East facade of reception hall (before renovation)

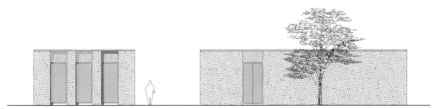

接待厅南立面图（改造后）/
South facade of reception hall (after renovation)

接待厅东立面图（改造后）/
East facade of reception hall (after renovation)

西安唐大明宫国家遗址公园御道广场　　西安

Emperor's Way Square of Xi'an Daming Palace National Heritage Park, Xi'an
2008～2010

建设机构 / Construction Organization：西安曲江大明宫遗址保护改造办公室 / Protection and Renovation Office of Daming Palace Heritage in Xi'an
项目地点 / Location：陕西西安 / Xi'an, Shaanxi
建筑面积 / Floor Area：232000m²
设计时间 / Design：2008年11月 / Nov.,2008
竣工时间 / Completion：2010年9月 / Sep., 2010

大明宫——盛唐时代的帝国象征，跨越千年的历史沧桑，如今只是一处历史遗存，含元殿基台不甚完整的土堆和略显乏力的保护性挡土砖墙显示着历史的沧桑和记忆的模糊。在这样的历史环境下我们考虑的不仅是我们拥有的能力，而是我们应该做什么？面对着历史和未来，我们绝不能采取有些城市在建设过程中忽视文物存在，割裂历史延续的做法。必须对其历史遗产进行有效的继承，实质性的尊重。只有在承上启下的可持续发展的过程中才能找到我们现代人的自尊，才有我们现代人的作为。

历史西安，唐城的意象已远，但她光辉、伟大与多元的文化内涵，仍深刻地影响着现代西安与未来西安。但是，没有人能够真实地再现盛唐的繁荣，也没有人能够确切地描述西安的未来，这一巨大的想象空间和时空跨越，给我们以设计灵感。

界面——一个用先进的技术手段构建起来的御道广场，可以将西安几千年积淀下来的灿烂文明之光折射到未来，将历史与未来连接在一起，这是我们进行大明宫御道设计的思考原点和始终坚持的创意。这个界面越纯粹，折射到未来的光芒就越深远，西安的发展空间就越广阔。这一"界面"是一个鲜活的场景，而非一块生动僵死的广场；具有生动丰富的内容，而非一成不变的形象。界面的构成基于真实、完整、可逆和全面保护的基本原则，它的意义在于以下几个方面：

1. "保护界面"：通过现状土壤的找平回填，采用新的技术手段，并设置隔水层有组织排水，排除雨水对下层土壤的影响。它作为"悬浮状态的工程体"覆盖在唐地面及其保护土层之上，绝大部分的遗址被保护了起来，同时我们选择了一部分遗址以"地面博物馆"的方式保护并展示。

2. "城市界面"：御道广场与含元殿、丹凤门形成一个宏大、开阔的"殿庭"。巨大的空间尺度和纯粹的地面铺装，将历史遗迹的沧桑感与现代城市公共生活有机地统一为一个整体，在促使人们体验历史的同时感受到城市公共生活的氛围。无疑这里将成为展示城市魅力和反映当代人文智慧的理想场所。

3. "文化界面"：丹凤门的复建，御道广场、历史遗存展示以及城市家居古典元素的烘托，使得含元殿遗址显得更加沧桑与震撼，他们共同提供了展示历史文化信息的平台，同时现代城市公共空间理念的注入将这里共建成为一道城市文化景观。

4. "生活界面"：作为现代城市广场，必须满足公众多样性的需求。"生活界面"提供一个全面

卫星影像下的大明宫主要遗址点 / Main heritage points of Daming Palace in satellite photography

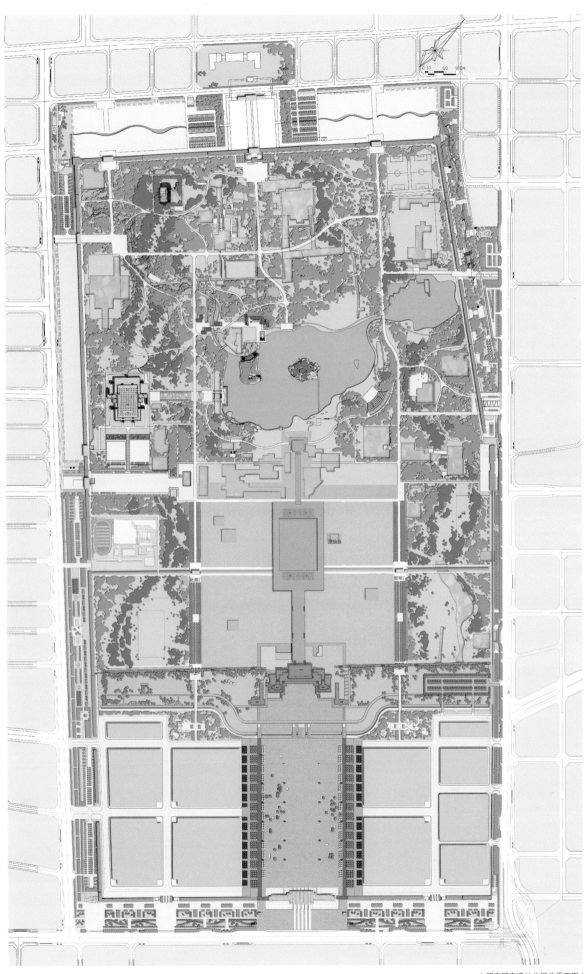

大明宫国家遗址公园总平面图 /
Site plan of Daming Palace National Heritage Park

的解决方案,将各种城市服务功能安排到外形统一的可移动的"城市家具"之中。可以根据广场使用需要,灵活组织"城市家具"的摆放。

5. "娱乐界面":在广场中央开敞部分随机设置一些LED地灯,公众在入夜后游览广场的时候会随意踩亮这些地灯,产生意外的惊喜,使广场夜间活动可以呈现出娱乐化的效果。

广场设计

御道广场在历史上是大明宫的"殿庭",如前所述,该区域在唐代是皇帝举行大典、阅兵、接见外国使臣的开阔庭院。其长度可以确定为现状含元殿与丹凤门遗址之间的开阔空间,其宽度则至今尚无定论,根据获国家文物局批准的《西安唐大明宫国家遗址保护展示示范园区暨遗址公园总体规划》,本次设计范围为南北长630米,宽360米的以含元殿至丹凤门中轴对称的区域。这个范围的空间尺度可通过与几个典型纪念性广场的对比来初步把握。

"界面"是以纯平广场为设计基本理念、通过6米×6米的"基本单元"展开的,整个广场统一在一个网格系统中,系统进一步细分不同的二级尺度和二级子系统。中部与含元殿约等宽的210米为开阔的庭院空间,以开阔、宏大的气势求得与历史记载的环境特质一致。在此范围之外,广场两侧发展为宽75米,长438米的树阵。在树阵中设计了宽18米的步行道,与原"上朝路"的位置相吻合,但宽度上作了较大幅度的加宽。步行道南北两端分别通向"龙首渠"遗址和"丹凤门广场",东西方向向两侧服务区过渡,使御道广场与相邻空间得以联系。

在考古发掘有唐地面或其他发现的地段,暴露考古回填保护的原生地面(也可在可靠的保护措施具备的前提下对遗址进行展示),充分展示考古遗迹,形成系列地面博物馆。广场南段及中段经考古发掘证明存在唐代历史遗迹的部分,北端作为考古遗址相对集中的区域,包括:"下马桥"、"上朝路"、"龙首渠"等考古发掘遗址。选取局部进行原址揭示展示,大部分遗迹则以回填土加以保护,对此两种情况,均在其四周加以围挡标示其位置,留给观光者充分的历史信息及想象空间。

丹凤门遗址鸟瞰 / Bird-eye's view of Danfeng Gate relics

沟渠遗址 / Relics of channels

上朝路遗址 / Relics of Shangchao Road

下马桥遗址 / Relics of Xiama Bridge

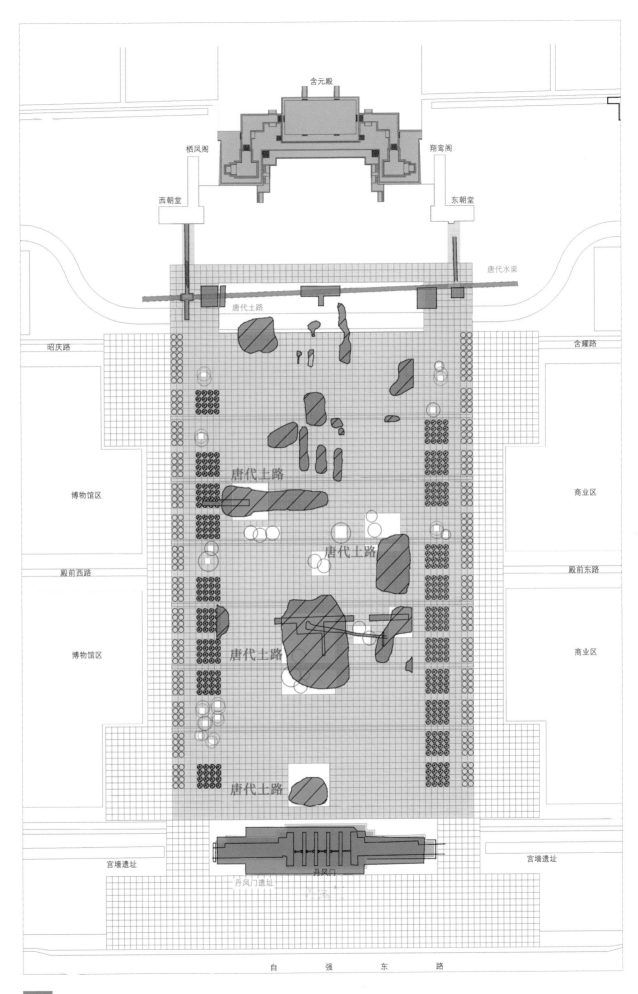

广场内遗址分布图 / Heritages distribution in sqaure

御道广场鸟瞰夜景 / Emperor's Way Square in night

御道广场的主要人流来自于丹凤门东西两侧宫墙开口，该出入口也是整个遗址公园的主要出入口，御道广场的交通组织以步行人流及电瓶车流线为主，丹凤门东西两侧服务区为人流集散广场，拟在此空间范围提供购票、旅游纪念品、团队集合、乘电瓶车等服务功能。电瓶车流线仅限于7米宽的建福东路和望仙西路，并在御道广场南端横穿并形成环线；人们步行流线必要时则可沿御道广场东西两侧树阵空间中行进，在非管控的时间段，广场绝大部分时间段是向所有步行者全面开放的，这是现代城市公共广场必然呈现的特质。

沿广场中轴线向东西两侧设置退晕布局的LED地灯，这些随机布局的地灯将在夜晚活跃整个广场的气氛，给游客留下难忘的记忆。同时发光LED地面的设计也考虑到对人的引导作用，LED的数量由两侧向中轴线逐渐加密，提示人们向着含元殿前的视觉展示区及丹凤门遗址博物馆聚拢。

在御道广场的设计过程中以上几项设计的深化设计都会面临不同主管部门的意见，但是界面的材质及技术做法是被最多次讨论，最终被确认为成"像砂土一样颜色和质感的"透水混凝土（又称露骨料混凝土），这是一种在国内尚未被广泛使用的具有透水性能的材料，大面积使用这种材料必然要面临许多技术上的问题甚至是观念上的问题。我们的设计团队克服了众多困难，一直跟进工程施工的全过程，但我们的初衷不变：那就是在整体保护"唐土层"不被破坏的前提下，通过采用最先进的技术手段构建起御道广场，实现文明的"叠加"，将西安几千年积淀下来的灿烂文明之光折射到未来，将历史与未来连接在一起。

城市家居 + 广场照明设计

广场作为城市中最为开阔的公共空间，在这里应该面向公众提供公共服务设施同时，基于公共安全考虑夜间还应提供必要的安全照度，因此理论上讲必须有高度超过40米以上的灯杆照明，若简单的以此进行设计高大的灯杆将会影响整个遗址广场的天际线，同时深埋的基础若选择地点不当将会有可能破坏地下遗址。这样不得不重新考虑照明设计方案，基于对遗址的尊重和保护的原则，我们设计团队提出了全新综合性的解决方案。

首先提出可移动的城市家居设想，这个装置将提供为广场公共生活准备的主要服务设施，其中包括：售货亭、饮水点、音响设施、咖啡厅、卫生间、观测点、信息点、垃圾容器、休息座椅、公用电话等，可以考虑一个系列统一的外形。其次，广场照明灯

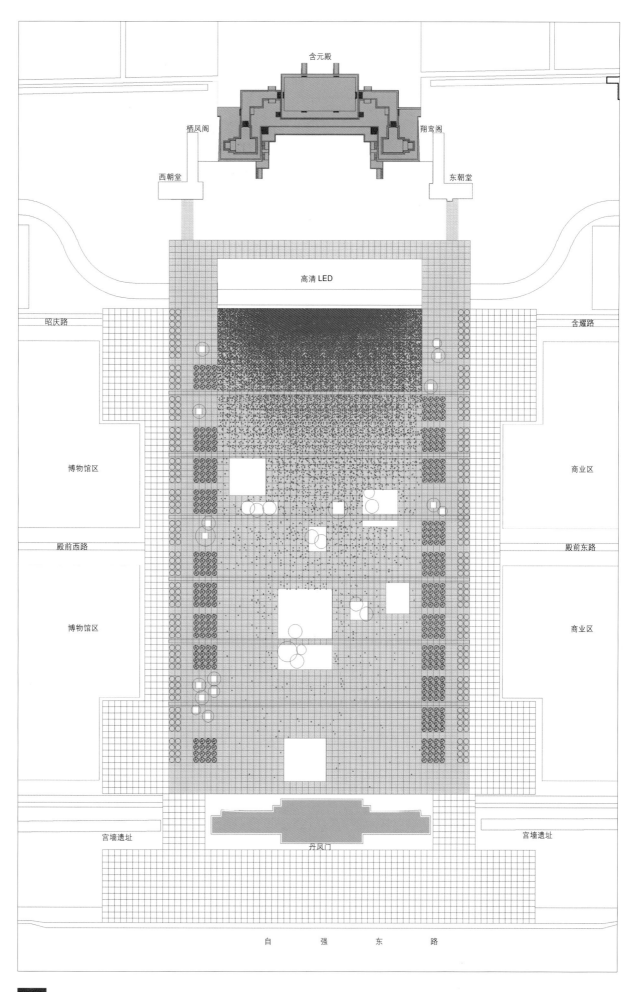

广场铺装及 LED 位置图 / Pavement of square and LED position

具的设计结合整个广场的设计理念进行创新。照明塔架与城市家具结合在一起,并采用反射式照明系统极大地降低照明灯具的高度,并可以在预设好的轨道上滑行,可根据需要调整广场照明灯具的位置和是否开启。这样的设计完全遵循"保护界面"的设计理念,不触动任何未知的考古区域并具有可逆性。装置本身是城市家具的组成部分,可以根据使用情况进行调整,既兼顾了照明的控制距离又可以使广场空间效果更具灵活性。白天照明灯具可以很好的隐蔽在门形现代装置之中,同时还可以为广场中部提供停留休息空间。这些城市家具将被设计得华美精致并焕发出古典主义的魅力。随着夜色的降临,灯具的机械装置徐徐展开,动态的美景,带给人们全新的体验。

为了实现上述设想使"城市家居"可以移动,我们设计了带滑轨的局部地面构造并结合雨水排放,设备电缆等相关要求,提出了浅埋综合管沟的

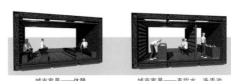

概念。综合管沟的设计尽量的减小埋深以避免对唐土层形成干扰和破坏。

绿化景观设计

根据历史记载及前人的研究,结合使用要求,殿庭绿化主要由国槐和银杏构成广场两侧的树阵。它有两大主功能:其一,强调并烘托出御道"殿庭"空间;其二,成为适宜的"生活界面"。通过绿化设计,结合绿化中的动态服务装置,使御道广场实现白天肃穆庄严、宏大空旷的氛围;夜晚带给人们亲切、繁华、时尚的感受。

在种植上尽量体现大气规整的种植肌理,采用单一纯粹的树种搭配,种植设计以常绿和落叶乔木为框架树种,为了防止冬季风对场地中人的影响,外围种植整齐的常绿乔木,主要指松柏类乔木,如油松,四季常青,既起到边框作用,也作为背景衬托重要景观。内侧种植落叶大乔木,达到浓荫蔽日的效果,冬季观枝,也能保证休憩空间充分的照明和日光。在经过进一步现场调查后,尽量保留有价值的现状树,并进行优化组合。

绿化景观的竖向设计关系到整体景观气势及空间氛围,在整体的景观设计中占有前提与基础的角色,并与建筑设计、植被设计紧密地关联,本方

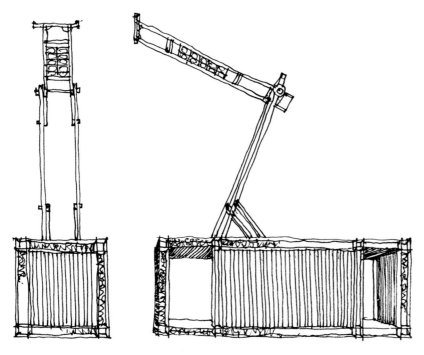

城市家具设计草图 / Design sketch of city furniture

城市家具夜景 / City furniture in night

案植被景观设计采用大量乔木，结合场地的实际情况——为保护古代土层不被破坏，使用了界面模数控制下的预制树池，种植竖向满足树阵种植的地形需要和自然的排水坡度并满足总体规划的基本标高控制要求。

技术手段与遗址保护

"保护界面"要求广场架在原地面之上，应承载起现代城市广场的需要，同时对唐土遗址层形成保护并具有可逆性。根据考古报告，唐地面距现地面约1米左右，在唐文化土层上应留300～400毫米原有土层作为保护，在此之上应运用可靠的技术手段实现遗址保护。我们的设计理念还包括用现代的技术塑造一个界面，这个界面越纯粹、越平整就越有深远的意义。占地面积约181200平方米纯平的御道广场，在遗址保护及技术要求上对设计团队提出了非常大的挑战。

首先考虑的是隔水设计，这是设计方案成立的最基础的环节，在原地表深300毫米内的原状土上堆填新土，重新规划场地竖向高程并在表层加入专用添加剂，充分搅拌并碾压压实，这种土体加固的方法可起到两方面的作用。一方面经过处理的土体具有良好的隔水性，能对唐文化土层起到保护作用；另一方面加固后的土体具有较高的承载力。

其次我们将整个场地，依托轨道分成若干个汇水区域，每个区域都贴临综合管沟或专用沟，排雨水主沟就置于其中，并向两侧找坡排入东西两侧的市政雨水管道内。这样就基本形成了有组织排水的基本格局。由于面层考虑土色的露骨料的透水混凝土，雨水会迅速下渗并在隔水层之上汇水然后排入支沟最终汇入主沟。在可透水的面层与下部的隔水层中间会有一道约500毫米厚的间层，这个间层不仅要求具有支撑结构的作用且要求水可以在其间自由流动同时还应该具备涨缩能力及易于施工的特点。基于以上的分析研究我们设计团队选择了级配碎石作为间层材料，关于碎石的级配比例也经过了多次的技术讨论，为保证可靠性要求施工单位针对该技术设计制作工程样板段。最终通过与甲方及施

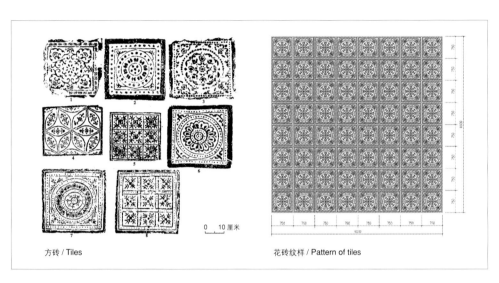

方砖 / Tiles

花砖纹样 / Pattern of tiles

工单位的反复论证将设计的原创性保存下来，这个开阔的广场即将呈现出异常平整的效果，现代技术和材料的应用在本工程遗址保护和利用实践中具有全新意义。希望现代人在感受厚重的历史氛围的同时也应该能体会到当代科技理念的些许光辉。

城市家具作为可以移动的机械装置本身具有为广场公共生活服务的作用，同时通过我们的设计使之具有历史感，体现出古典艺术的片段与当代科技的结合。精确的反射照明技术得到应用，通过电子程序我们可以模拟出广场任意位置的照度分布情况，反射板上下的镂空设计将极大地减小风荷载对其稳定性的影响，起降臂的液压系统在极小力臂情况下仍能完成反射照明系统的升降动作。城市家具上还安装着摄像监控系统，更加有利于为广场公共生活提供必要的服务。总之本设计的核心目的是遗址保护和利用，所有的现代技术手段和材料都应围绕这个目的而进行。

当今御道广场的位置属于考古学定义的"殿庭"位置，千年以前是一块空场，所以实物遗迹并不很多，但是仍然有部分内容需要展示。

例如：在今含元殿南遗址保护范围南侧考古发现了东西走向的唐代水渠道1条。该渠道东西贯穿整个拆迁范围，经钻探试掘而明确的渠道长度400余米。发掘得知，该渠道直接叠压于现代建筑堆积层之下。从发掘的基础断落来看，该渠道与含元殿南沿基本平行，宽约4米。从剖面观察，该渠道文化堆积自下而上大致有四层，依次为：紫红色胶泥质土，厚0.9～1.7米，内夹有些白色淤土，另外还有大量唐代砖、瓦、瓦当出土；黄色淤土，土质较硬，干燥厚0.8米；浅黄色淤土，质地松软细密，剖面显示有水锈斑，厚0.8～1.07米；青灰色淤泥，成块状，较黏硬，水锈斑重，厚约1.1米，应是原始渠道底浸渗而成。

在渠道内还出土有大量的唐代砖瓦、石块螺壳、陶瓷器、铜钱、铁钉、铁剑等遗物。对于这样的遗迹，我们设计采用覆土保护并在地面用砾石标识保护展示方式。

对于中央桥梁遗址的保护展示我们将采用地面博物馆的方式进行。和御道广场里的唐代路土展示方式一样，也是通过制造一个封闭的空间将遗迹保护起来，并展示给地面上的游人。在隔绝处理的前提下，利用设在旁边的升温恒湿设备控制遗址处的环境条件，以达到保护遗址的目的。

Daming Palace is the imperial symbol of glorious Tang Dynasty, and is now just an historic relic through the vicissitudes of history over a thousand years. Incomplete remnants and feeble protective brick retaining walls at the foundations of the Hanyuan Hall bear testament to the vicissitudes of times and the vagueness of memory. In such historical circumstances, we consider not only that we have the ability, but what should we do? In face of the history and future, we must not follow the practice that some cities ignore the existence of cultural relics in the construction process and separate the historical continuation. Historical heritages must be inherited effectively with substantive respect. Only in the continuous process of sustainable development can we find self-esteem and accomplish something in this process.

The image of Xi'an as a Tang city has been far, but her brilliant, great and diverse cultural connotations have still had a profound impact on modern Xi'an and future Xi'an. However, no one can faithfully reproduce the prosperity of the Tang Dynasty, and no one can exactly describe the future of Xi'an, so this huge imaginary space and spatiotemporal span give us design inspiration.

Interface-Emperor's Way Square, built up by advanced technical means, can reflect the brilliant civilization of Xi'an over thousands of years into the future, and connect the history and future together. This is the starting point in our thinking about the design of the Daming Emperor's Way Square and a creative idea that we have upheld throughout. The purer the interface, the further into the future the light will reflect, and the more development space Xi'an will have. The "interface" is a vivid scene rather than a rigid square; it has vivid and rich contents rather than rigid image. The composition of the interface is based on the basic principles of truth, completeness, reversibility and full protection, and its significance lies in the following aspects:

1. "Conservation Interface": By leveling and backfilling the existing soil, adopting new technological means and setting the aquiclude for organized drainage, it can exclude the impact of rainwater on the subsoil. As a "floating structural body" lying over the Tang ground level and its protective layer of soil, it can conserve most of the relics. We have also chosen a part of the relics to protect and show in the way of "ground museum".

2. "Urban Interface": Emperor's Way Square, the Hanyuan Hall and the Danfeng Gate form a huge and open "palace courtyard". The vast space and pure ground pavement have integrated the vicissitudes of historical sites and modern city life as a whole, to encourage people to experience history and feel the atmosphere of city public life. No doubt there will be the ideal venue to show the charm of the city and reflect contemporary human wisdom.

3. "Cultural Interface": The rehabilitation of Danfeng Gate, the demonstration of Emperor's Way Square and historical relics and the rendering of classical elements for urban furniture make the Hanyuan Hall appear even more vicissitudinary and spectacular, and they together provide a platform for displaying historical and cultural information, while the concept injection of modern urban public space has built it a cultural landscape.

4. "Living Interface": As a modern city square, it must satisfy various needs of the public. The "Living Interface" provides a comprehensive solution, placing a wide variety of urban services into moveable "urban furniture" with a unified exterior. This "urban furniture" can be arranged flexibly according to usage requirements.

5. "Leisure Interface": LED floor lights are randomly installed throughout the open area in the center of the square. Visitors to the square turn

2010.1.14

2010.3.18

2010.3.18

建设过程 / Construction progress

2010.4.7   2010.5.19   2010.6.17

these lights on by stepping on them, giving them a happy surprise and turning the square into a leisure venue.

Square Design

Emperor's Way Square in the history was the "palace courtyard" of Daming Palace. As mentioned earlier, the area in Tang Dynasty was the open courtyard where the emperor held a ceremony, parade and meeting with foreign envoys. Its length can be determined to be the open space between the existing Hanyuan Hall and Danfeng Gate, and its width is so far inconclusive. According to the "General Plan for the Xi'an Tang Daming Palace National Historic Site Model Display Area and Site Park" approved by the State Administration of Cultural Heritage, the design range from north to south is 630m long and 360m wide, flanked by the Hanyuan Hall and Danfeng Gate. One can get a basic understanding of the size of this space by comparing it with several classic commemorative squares.

"Interface" makes a flat square as the basic design concept and develops from the 6mX6m "basic unit". The entire square is unified in a grid system which is further broken down into secondary scales and subsystems. The central part (210m) with approximately equivalent width to Hanyuan Hall is the open courtyard space, which strives to match the environment described in historical records with its openness and grandeur. Outside this range, both sides of the square have been developed into tree array of 75m wide and 438m long. 18m-wide walkway has been designed in the tree array, which is consistent with the position of original "Upward Road", but substantially widened. North and south ends of the walkway lead to the "Longshou Ditch" site and "Danfeng Gate Square", and the east-west direction has transferred to both sides of the service area, which makes Emperor's Way Square connected to the adjacent space.

The Tang grounds or other lots discovered during archaeological excavation have exposed native ground for archaeological backfill protection (the site is shown with the premise of reliable protection measures), fully displayed the archaeological remains and formed a series of ground museum. Archaeological excavation has proved the existence of the Tang Dynasty historical sites in the southern and middle sections of the square; the northern end as a relatively concentrated area of archaeological sites includes: "Xiama Bridge", "Upward Road", "Longshou Ditch" and other archaeological excavation sites. Some parts are selected to reveal and show its original sites, and most of the ruins are backfilled for protection; for these two cases, enclosure is made all around to mark its position, leaving the tourists full of historical information and imagination.

The major pedestrian flow of Emperor's Way Square is from the palace wall opening at east and west sides of the Danfeng Gate, and the entrance is also the main entrance of the entire heritage park. Traffic organization in Emperor's Way Square is mainly based on walking pedestrian flow and battery cars. The service area at east and west sides of Danfeng Gate, as the assembling and dispersing square for the pedestrian flow, intends to provide tickets, souvenirs, group assembly, taking battery car and other service functions. Battery car is limited to the 7m-wide Jianfu East Road and Wangxian West Road, and also travel across the southern end of Emperor's Way Square to form a circle line. Walking pedestrian flow line, if necessary, can advance along the tree array space at east and west sides of Emperor's Way Square; in the non-control time period, the square is fully open to all walkers in most of the time period, showing the inevitable characteristics of the public square in modern city.

LED floor lights with color change layout are set along the square axis to the east and west sides, and these random layout of lights will activate the whole atmosphere of the square at night, leaving an unforgettable memory to tourists. The design of flashing LED ground has also considered its guiding role for the people, and the number of LEDs has gradually increased from both sides to the axis, prompting people to gather toward the visual display area in front of the Hanyuan Hall and Danfeng Gate Site Museum.

During the design process of Emperor's Way Square, different departments in charge have different opinions about the detailed design for the above items, and the interface material and technical practices have been discussed for the most times, and was eventually identified as pervious concrete like "the sand color and texture" (also known as exposed aggregated concrete), which is a kind of permeable material not yet widely used domestically, and a wide use of this material will inevitably be faced with many technical and even conceptual problems. Our design team has overcome many difficulties and followed up the whole construction process. However, our original vision has remained unchanged: providing that the whole "Tang surface layer" was protected from damage, we use the most advanced techniques and methods to construct Emperor's Way Square, so as to realize an "overlapping of civilizations", reflect the light of Xi'an's several thousand years of glittering culture into the future and link the past and the future together.

Urban Furniture + Square Lighting Design

The Square as the most open public space in the city should provide public service facilities and necessary security illumination at night for public safety consideration, so in theory, there must be lamppost lightings with a height of more than 40m. However, high lamppost according to such a simple design will affect the skyline of the whole site square, and improper location of the deeply buried foundation will damage underground ruins. For this reason, we have to reconsider the lighting design, and based on the principle of respect for and good protection of the sites, our design team has proposed a new comprehensive solution.

First, the assumption of movable urban furniture is proposed, and this device will provide the main service facilities for public life on the square, including: kiosk, water point, audio facilities, cafe, toilet, observation point, information point, garbage container, rest seat and public phone, which can be considered with a unified exterior. Second, the design of the square's lighting brings together the design philosophy of the whole square as a basis for innovation. Lamp poles and urban furniture are combined, and a reflective lighting system greatly reduces the height of the lighting and can glide in the preset orbit; the location and status of the square's lighting can be adjusted according to the needs. This design is in full compliance with the "conservation interface", with no unknown archeological disturbed, and it is reversible. The device itself is an integral part of the urban furniture, and can be adjusted based on usage, thus satisfying the lighting control range while providing more flexibility in spatial effect. During the day, lighting can be well concealed in the door-shaped modern devices, and can also provide a space to stop and relax in the center of the square. The urban furniture will be ornate and delicate, exuding a classical charm. With the advent of nightfall, the lighting mechanism slowly unfolds, and the dynamic spectacle gives people an entirely new experience.

To realize the above assumptions and make "urban furniture" movable, we have designed local underside structure with rails and combined it with drainage, equipment, cables and other related requirements to propose the concept of comprehensive shallow-buried pipe ditch. The design of comprehensive ditch minimizes the depth to avoid interference and damage to the Tang soil layer.

Design of Landscaping

Based on historical records, former researches and application requirements, the palace courtyard landscaping is mainly composed of Chinese scholar trees and ginkgoes which form tree arrays at both sides of the square. It has two major functions: the first is to highlight and render a Yudao "palace courtyard" and the second one is to become a suitable "living interface". By means of landscaping design and combining dynamic service devices in landscaping, the Emperor's Way Square is quite solemn, dignified and spacious in the daytime while amiable, prosperous and modern at night.

Planting is designed to embody generous and neat planting texture as much as possible by applying only one kind of tree species for collocation. Thus, evergreen and deciduous arbors are adopted as framework tree species in planting design. In addition, to prevent the negative influence of winter wind on people in the site, orderly evergreen arbors are planted around, mainly referring to conifers trees and Chinese pines which remain green throughout the year and can server as both the framework and the background to set off important landscape sceneries. Deciduous big arbors are planted inside of the square in order to form lush shade, block out the sun and guarantee adequate lighting and sunshine in leisure space while viewing branches in winter. After further onsite survey, we decided to preserve existing valuable trees as much as possible and conduct optimization.

Vertical design of the landscape design involves the overall landscape momentum and space atmosphere. It is the premise and foundation in the overall landscape design and is closely connected with architectural design and vegetation design. In the vegetation landscape design of this plan, a large number of arbors are adopted. Considering the practical situation of the site, we applied preset tree grates under the interface module control to protect the ancient soil from damage, and vertical planting to meet terrain requirements of tree array planting and natural drainage slope as well as basic elevation control requirements of the general plan.

Technical Means and Historic Sites Protection

"Conservation Interface" requires that the square be located on the original site and bear the needs of modern city square. In the meantime, it should be protective to Tang relics soil layer and be reversible. According to archaeological report, the Tang ground is about 1m high from the existing ground, thus, 300mm to 400mm original soil layer should be reserved on the Tang cultural layer as protection, what's more, reliable technological means should be applied to realize the conservation of the historic sites. Our design concept also includes using modern technology to create an interface. The purer and the neater the interface is, the more significant it will be. Emperor's Way Square is about 181,200m^2, which poses great challenges for the design team in terms of historic sites protection and technical requirements.

The first consideration is waterproofing design which is also the fundamental process establishing the design scheme. We placed new soil on undisturbed soil which is 300mm deep inside of original ground level, and also redesigned vertical elevation of the site and added dedicated additive to the surface layer. The additive was then fully stirred and rolled for compaction. Such a method of soil reinforcement has two functions: the first one is that the processed soil can separate water very well so that it can be protective to the

2010.8.8

Tang cultural layer, and another one is that the reinforced soil has higher bearing capacity.

Then, we divided the whole site into several catchment areas in accordance with tracks and each area is adjacent to the comprehensive pipe ditches or dedicated ditches; besides, the main rainwater sewer is located inside of the catchment area and discharge water into the municipal drains at both east and west sides via slopes. In this way, the basic pattern of organized drainage is established. Since the pervious concrete of exposed aggregate which considers earth tone on the surface and the rainwater will infiltrate rapidly, gather above water-resisting layer and flow into main ditch via lateral ditch, a 500mm thick interlayer is provided between the permeable surface layer and the lower water-resisting layer. The interlayer is required not only to serve as the supporting structure, but also to allow free water flow within it, expansion and easy construction.

Based on the above analysis and study, our design team chose graded gravel as interlayer materiel at a gradation ratio determined after repeated technical discussions. For this technology, the construction unit is asked to design the engineering model to ensure its reliability. Eventually, the originality of the design is maintained after the repeated demonstration by party A and the construction unit. This spacious square will soon present its extreme neatness. The application of modern technologies and materials presents a totally new meaning in the historic site conservation and practical utilization. It is hoped that the modern people can experience the splendor of contemporary science and technology concept while feeling the profound historic atmosphere.

The urban furniture itself, as movable mechanical devices, can serve for public square life and at the same time, with our design it can bear a sense of history where we can see segment of classical art combined with modern technology. Precise reflective lighting technology is applied, and we can simulate lighting layout at any position of the square through electronic program. The hollow-out design above and below the baffle board will largely reduce the negative influence of wind load on it. The hydraulic system of lifting arm can lift the lighting system even as the lifting arm force is very low. Camera monitoring system is also installed on the urban furniture to provide better service necessary to the square public life. Above all, the main purpose of this design is to protect the historic sites and the application of all modern technologies and materials should serve for the purpose.

The present Emperor's Way Square is situated on the "palace courtyard" defined by archaeology. However, it was an empty space thousands of years ago, so there are not much relics left, but still some parts to be shown.

For instance, an east-west water channel in Tang dynasty at the south side of Hanyuan Hall south conservation area is discovered by archaeologist. This water channel goes through the entire range of demolition, and by a drilling test dig, it is found as long as 400m and is directly overlapped under present building accumulation layer. Observed from the foundation fragmentation of the water channel, it is almost parallel with Hanyuan Hall and about 4m wide. Observed from the cross section, the water channel cultural accumulation comprises four layers from top to bottom:

Amaranth daub soil, 0.9m to 1.7m, inside is white soil; additionally, a considerable quantity of Tang bricks, tiles and eaves tiles are unearthed. Yellow harder soil, 0.8m thick when dry. Light yellow soil, soft and fine, with water rusty spots on cross section, 0.8m to 1.07m thick. Blue gray soil in bulk, much sticky stiff and lots of water rusty spots, about 1.1m thick; this could be formed as a result of water penetration at original channel bottom.

There are a large number of Tang tiles, stone shells, pottery and porcelains, copper coins, iron nails, iron swords and other relics inside the water channel. For such relics, we planned to adopt soil cover conservation and use gravel mark for protection and exhibition.

As for the conservation and exhibition of the center bridge relics, we applied ground museum. It is the same as Tang road soil exhibition in Emperor's Way Square by creating an enclosed space to protect the relics and to exhibit them for visitors. On the premise of enclosed process, a warming and constant humidity device at one side is used to control the environmental conditions around the relics for the purpose of protecting the relics.

In a word, if the seven main principles set forth in Charter for the Interpretation and Presentation of Cultural Heritage Sites (Draft) (2007) by International Council on Monuments and Sites (ICOMOS) are used to assess the previous work, it can be said that the "interface" is helpful for the public to know cultural heritage sites materially and culturally. Thanks to our predecessors' effort, we can base our work on the evidences collected by rational scientific and academic methods as well as on evidences collected from life, culture and tradition. By restoring the Hanyuan Hall and Danfeng Gate sites as well as the interface above the protective soil layer, the exhibition of Emperor's Way Square can be connected more closely with its extensive social, cultural, historical and natural backgrounds; what's more, all of these are in accordance with the fundamental principles of authenticity set forth in the Nara Document on Authenticity (1994). Based on the consideration for the future, the "interface" will adapt itself to its natural and cultural backgrounds and be socially, economically and environmentally sustainable.

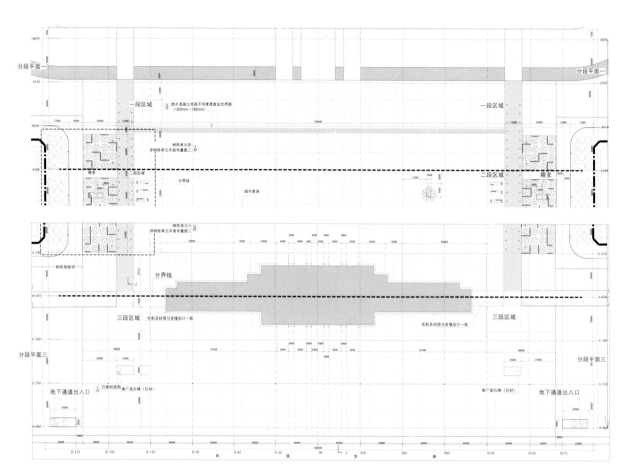

三段平面图 / Plan of Part 3

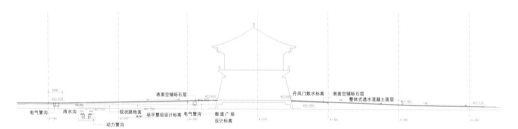

1-1 剖面图 / Section 1-1

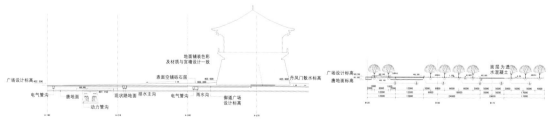

2-2 剖面图 / Section 2-2   3-3 剖面图 / Section 3-3

一三段平面、剖面 / Plan and section of Part 1 and Part 3

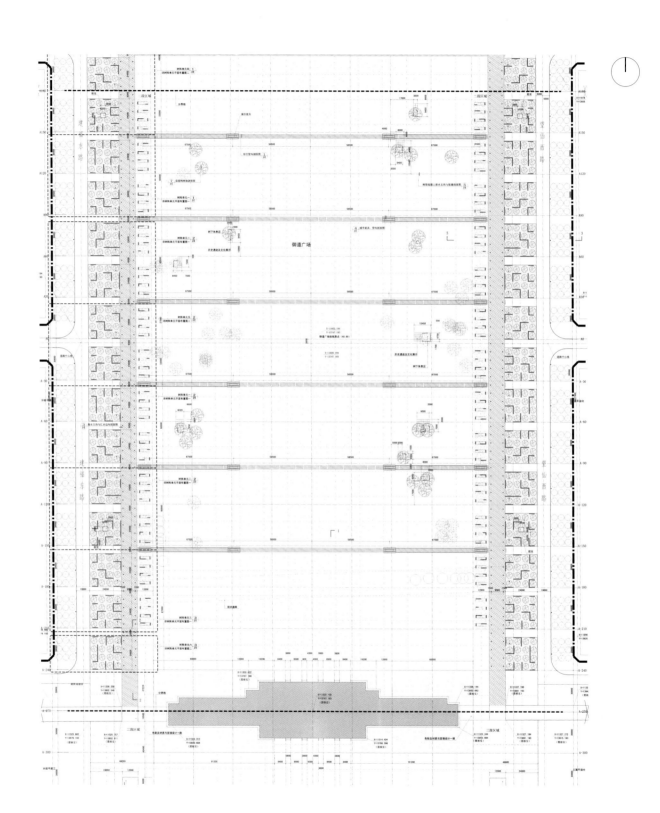

二段平面 / Plan of Part 2

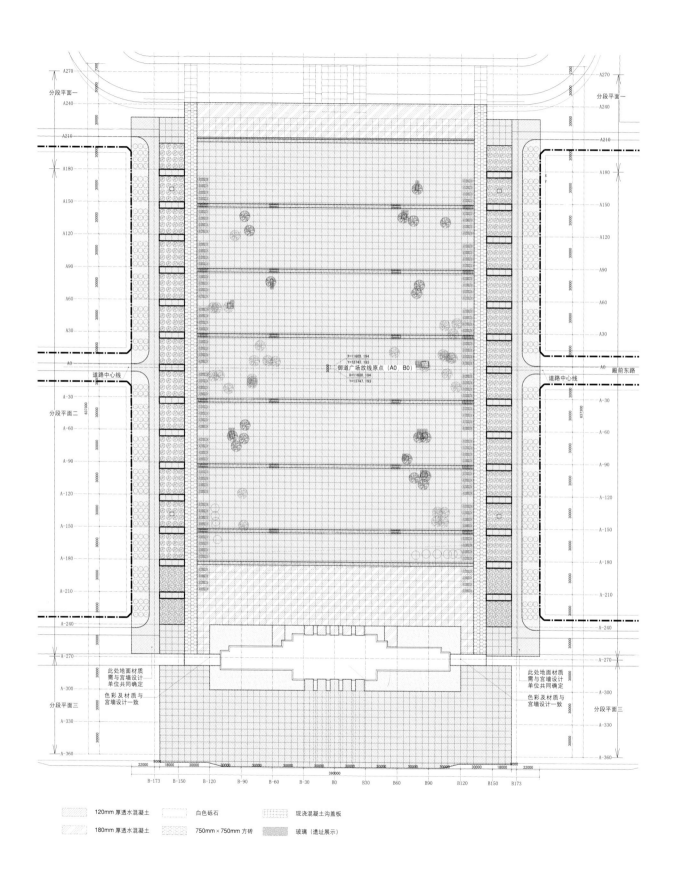

广场场地铺装分布图 / Pavement of square

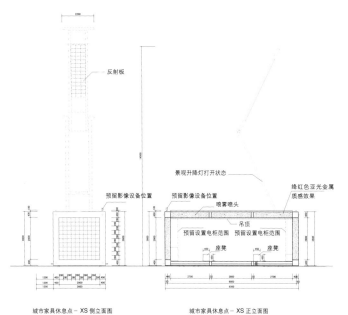

城市家具详图 / Details of city furniture

广场铺装与管沟连接处剖面 / Section of square pavement and channel joint

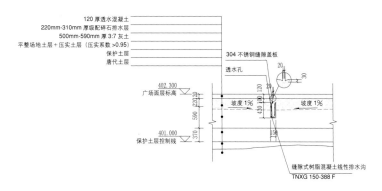

树脂混凝土排水支沟与膨胀剖面图 / Drainage channel of acrylic concrete and expanded section

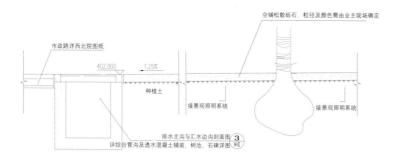

国槐树池剖面图二

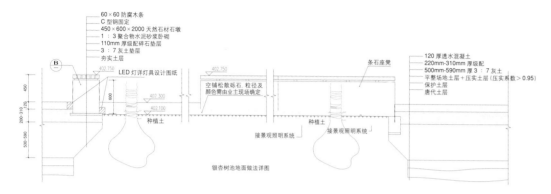

银杏树池地面做法详图

树阵设施及局部大样 / Details of facilities of tree grid

广场1号遗址展示坑顶平面大样图

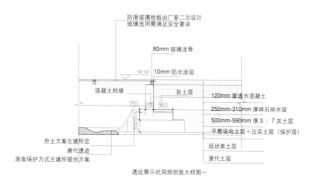

遗址展示坑局部剖面大样图一

遗迹及历史文化展示平面、剖面大样 / Details of heritages and historical culture exhibition

朱小地访谈 / Interview

采访人／黄元炤
北京 2011.10.12

黄：您进北京院后，曾被派往海南分院工作，而海口寰岛（泰得）大酒店是您在海南时期的代表性作品。这个项目给我两个印象：一个是由于海南的植被生长快，植物繁多，是热带雨林、热带季雨林的原生地，所以，可以观察到您对自然有所设计，建筑室内有椰树，有海景观光电梯，且绿化覆盖率达30%，充分体现海南地区的偏向于热带的特色，而似乎从这个项目开始，您将植物运用到建筑之中，比如后来的"秀"吧与"旬"会所。一个是建筑形体部分，由于海南常年光照充足，光合潜力高，所以，您设计了一个高达27米的四棱锥网架结构大堂，给室内提供充足的光线与亮度，而直指蓝天的三角塔形尖顶，像是西方古典教堂钟楼或是中国古典塔楼的感觉，带有点抽象化的符号象征意义，能谈谈这个项目吗？这部分您的看法如何？

朱：对于自己的设计，我有这样的感觉，建筑建成后，很少有摄影师能将房子的实地感受表达出来。我想原因主要在于：我的设计希望环境能跟人结合，而不是将建筑和人对立起来。我希望营造一个场所、一种氛围，把人放到那个场所和氛围中去，并让身在其中的人们体验和感受。我认为在建筑设计中，建筑师并非主角，相反，身在其中的观者、使用者才是主要的，只有你设计的建筑与环境影响到人，使人产生感受，人才能注意到建筑设计的初衷，才能形成真正的对话关系。

基于此，我的设计有时候会让人感到有点"碎"，我擅长从某一点开始，追求某一种氛围，并让其充斥在环境中。相反，我不太习惯于西方建筑师的设计方式，比如盖里、扎哈的设计，或者我还没有掌握那么一种能力，那种无中生有、通过简单的塑形去影响别人的能力。所以，我做酒吧这类房子，还有点路子，也比较在行；要让我做那种很夸张的体型，并强迫用材料去表达这样的造型，我找不到感觉。

黄：盖里与扎哈的设计，夸张的体型，通常属于个性化的那种。我想，这种对于设计的掌握与无法掌握，放在现实层面上，很难有定论。

朱：设计上个性化的东西，可能跟每个建筑师个人的价值观有关。我从小在一种正统的教育中成长，中学里是第一批团员、大学中做班级的学习委员，集体观念在这样的成长过程中变得根深蒂固，擅长并习惯于以一种群体意识来处理问题。我发现用这种方式，很容易跟我的客户以及周围人建立一个比较好的社会关系，这也可能是我的一个优势。

黄：您刚提到希望有个场所，去营造一个氛围，让别人去感受，没错，就我观察您的作品，确实有要塑造一个氛围。比如说寰岛泰得大酒店，您企图要塑造一种热带的氛围，这是一种空间氛围的设计倾向。比如说银泰中心顶层的"秀"吧也有点类似，有清澈的水池或水膜、有树、有植物、有音乐、有暖黄的光与间接的藏光，也是要塑造一种时尚、休闲、轻松的氛围。所以，氛围是您设计中的重点，但"秀"吧的平面布局还是从功能考虑的，带有点中国传统序列空间的轴线布局与方法。而当看到"秀"吧的宋式大屋顶出现后，中国古典的氛围又涌然而生，仿佛是传统官方宅第或乡绅员外宅第的现代再现，加上情调空间氛围的营造，您似乎想在传统建筑形态中体现现代时尚的氛围，这部分您的看法如何？

朱：我这些年的建筑设计中，对几个方面的问题做了研究。

一是城市空间的研究。城市不是一个简单的专业。城市的比较研究、城市经验的获取，这是我对城市研究的部分内容。其中，动态规划，强调的是公共利益，对应的是公共空间，公共空间具化为一个自由开放的步行系统。对这方面，我做了新的理论阐述。

二是传统建筑空间的研究。我对传统空间有几点理解：首先是方位感。中国传统建筑的特点是建筑与礼制结合在一起，中西方建筑在这一点上有明显的分别。中国建筑空间强调方位感，有中轴线、左右对称，由此产生的层次、层级，并建立一种从低到高的对应关系。人一旦进入这个建筑空间中，会立刻意识到自己的位置，意识到自己在这个建筑中处于怎样的关系中。方位感是中国建筑的一个特点，这个特点在西方建筑中很难找到。因此，我特别喜欢用一些轴线关系，用一些明确的空间提示。此外，在我的建筑设计中希望达到的程度是没有室内和室外之分。我习惯于从环境的角度和尺度切入设计，先思考院落格局，然后是建筑，然后要跟室内空间发生关系；我所理解的建筑，是通过多重的空间关系，从室外到室内不断演进的过程。于是，必须先注重环境，特别是院落的关系，注重"院"在建筑中的位置。

三是建筑空间层次的研究。空间的层次、递进、反复演进的过程，与我提出的"层的编辑"的设计方法完全对应。你所提到的"氛围"，实际上是一种场景感。好的建筑是可以对人产生影响的，那身在其中的人如何被影响了呢？我的方法是利用"层"。什么是"层"，比如一个建筑立面，可以分成若干层，每一层可能是同一种材质，也可能是不同的材质，并且形成一定的分隔比例，这种层次可以达到一定的秩序感。一个立面我们可以这样来理解，相对的，一个空间也如此。

人在现代建筑中的感受是什么，我的理解，就是穿越层次的过程，以及所得到的感受。你不要把现代建筑理解成为一个"房子"，"房子"只是工程师和施工人员围起来的东西。你要体会一层层的过渡中穿越层次的感觉，这种体验是一个不断向前的过程，与现代人的生活一致。而建筑师追求的目标是什么呢？就是人在空间中停留的意义。不是简单的留点沙发家具，而是要启发人的联想、感受和思想，他的思想或心灵要跟建筑空间有所对话。这样的对话关系，就是"氛围"。从我个人的创作体验来看，这种"氛围"的塑造，不是一件事重复了很多次，而是散点的。

黄：您说到的散点，我有注意到，您都用一些比较零碎的物件去构成所谓的场所与氛围。

朱：这就是设计需要考虑的东西，能不能将零碎的物件整理起来，让别人进入建筑的时候，也能够按照你的心灵轨迹来体验；而不是仅仅将建筑方法、形式的东西搞得很纯粹。我觉得这种"散点"的概念，这种相互搭配的影响关系，是很有价值的，也是我在设计中经常使用的。尽管最终实现的建筑还不多，建筑设计中也使用了其他的处理方法，什么扭转、叠加、材料的相互对比等等，但这些实现的作品中若能达到这种效果，都还是有这种关系的存在。

建筑师在这里要习惯于编辑这个"层次"，不同的方向与角度该如何考虑，如何反复寻找转向的可能性来对应"层"的关系。这种设计是要慢慢进行的、一点点体验的，因此非常消耗精力，它没有室内室外之分，建筑的各个角落都要考虑到，因此，设计变得很复杂。思想过程中的亮点、好的东西，要把它们串在一起，你想的很多很累，但关键的是，你要能把它们整理好，有取舍，这很难。

黄："秀"吧，白天去跟晚上去的感觉是不一样的。白天去的时候，可以体会到经由这个节点转到那个节点，看到另外一个场景，有种步移景异的效果。然后看到宋式大屋顶后，当时现场的记忆被拉回到久远古老的年代，仿佛游走在宫城的大街上或是穿行在大院宅第之间。近观宋式屋顶的形态结构之美，或是远观旁楼的规矩高耸之律，体会到与天空接近的感觉；可是到了晚上，音乐一放，灯光一打，吧里的爵士乐舞曲响起来，各种酒味与庭院里的休闲座椅，是一种在视觉、听觉与味觉都处于非常时尚的氛围，是很现代的。所以，白天与晚上的"秀"吧感觉是截然不同的，这让我产生非常强烈的冲突感，记忆被拉回去历史后，过不久，在视觉、听觉与味觉的部分又如此的现代。由此可知，您的设计总有一种旧与新，传统跟现代的冲突与结合的关系，就像您说您想打破室内与室外的界线，而传统和现代的界线是不是也是您想打破的，或者更是一种旧与新的融合，这部分您的看法如何？

朱：你说的"打破传统和现代的界限"，我的理解是，怎样把建筑的时间轴作为设计的主轴，让人在这个地方能待得住，有停留的意义。在确立这个主轴后，建筑的形式设计就可以放松到最简单的状态。我觉得除了"秀"吧外，我的其他设计方案，都存在旧与新、传统与现代的冲突感。白天去时会怅然若失，但到了晚上，尤其有活动的时候，你会发现各种现代元素都跑出来了，新、旧元素在时间轴上又找到了重新集合的机会，感觉非常奇妙。

黄：所以，您的设计会先强调轴线关系，创造出一条似真似假的路径，有着不同的场景，加上一进一进的层次逻辑，最用散点的方式营造出特有的空间氛围。这当中会有很多转折的过程，比如说我看您另外一个作品——"旬"会所，进去时会先经过一个钢构的大门，这也是一种转折。

朱："旬"会所的门，相当于四合院宅邸的正门，人并不从这里进入，而是往两边走，从边上绕。这个东西，说到底有什么设计呢？其实就是那么一点点设计，就是把中国四合院先抑后扬的感受传达出来。

黄：我又观察到在"旬"会所和"秀"吧，您对于水的应用特别明显。在"秀"吧，建筑是体与体的组合，中间有一处水池，旁边摆一棵树，您似乎想用水与植物创造出一种东方意境，是自然与人工的相互辉映。在"旬"会所中，水在旁边，有一个过道通过去，利用水区隔开内外之间的关系，加上几道墙与游廊组合起来的片段，创造出一种转折或是端景的路径或轴线。所以，您是不是特别想用水、气与植物去制造出一种中国特有的意境，将深远文化体现在现代空间之中，让视觉与感官产生一种仪式性的行为，您对于这两个项目怎么看待？

朱：水，一般在商业项目中都会用。生意场上，水就是财，我们的客户一般都喜欢。实际上你刚才讲的水，是时间的引子。水是流动的，可以让你意识到时间在这个过程中的作用。然后水和灯光、和焰火、和激光结合起来效果会更好。我总是强调，你可以透过这种景色想到过去、也可以想到未来；时间轴非常重要。这样的时间轴，贯穿了我的"城市收藏"的一系列作品，时间在建筑作品中的作用非常强。它无法通过摄影镜头来捕捉，平面图片是无法表达的，即使我自己去拍照也是徒劳，它需要亲身体验。

黄：在"旬"会所中，可以看到一些古建筑的局部物件，或是古建筑物件元素藏于玻璃盒子之中。就我的了解，您在清华读书期间，曾拾到一块砖雕一直收藏至今，这勾起了您收集传统建筑构件的兴趣。1998年，您到山西等地区考察古建筑遗址后，意识到必须要认真学习传统文化。后来，您担任院长后促进了中国文物局与北京院共同组织"重走梁思成古建筑之路"等传统文化考察活动。您对传统文化的体会与认识，从一个点到一个面，再扩展到形成一个机制，能谈谈这部分的转折吗？能谈谈在收集古建筑构件的经历与经验对您个人的设计思想有什么样的启发？而这些收集来的古建筑物件，是否会去考究它？

朱：已经很多年没有收藏古建筑构件了。我强烈地感到一个很有意思的事情，如果你在收藏中有了感悟，那么你到了一个真正的传统建筑环境中，就会感到传统的存在，而且会去细细地观察、品味它，进而感悟它。但只要你一离开这个环境，就没有了这种感觉与感悟，即使当时当场你看得很清楚，即使你手中拿着当时的照片，也很难体会那个环境给你的感受。你会发现传统的东西，就是一个氛围、一个环境。也就是说，中国建筑就是人和建筑的二元关系，没有人就没有建筑，建筑就是为了人而存在的。因此没必要简单地强调某一方，你强调人或强调建筑，其实没有意义，因为实际上这是一个二元的关系。这是我对传统建筑的一个理解，这种对二元关系的理解，也应用在我的设计实践中。我希望形式上表现得尽量放松、尽量简化，但人还是在建筑物中使用着它，比如SOHO现代城、中国石油天燃气集团公司总部大厦等项目，还有一些未建成作品，轮廓都非常简单，不需要过多的形式。另外，我所有的项目都有"院落"的概念，或是平面的，或是立体的，院落的概念与我所说的二元关系有着直接联系，那就是一旦这个建筑固定下来，就变成一个新的躯壳，不需要建筑师的表演，生活在其中的人们自有他们的精彩。所以我的设计可能和其他人不一样，我希望作品越干净越好。哈德门饭店中的立体四合院、中国石油天燃气集团公司总部大厦中几个框构成的立体的院，都是这个院落概念的演化。现在正在施工的深圳的一个高层建筑项目，南北各是一大片绿地，房子在中间，我的策略是将建筑首层的混凝土核心筒放在两边，剩下的墙体全部采用玻璃，南北相通，就是一个院子。这里什么都没有，就是一个大堂，进去后，感觉这么干净，室内外是一个共同的空间，我就是要做这个。

黄：您现在还有出去考察古建筑吗？

朱：很少了。前几年有名的地方，我都去过了。另外，就如我刚才说的，要真的考察，就要真的认认真真进行。这个我现在做不到了，比如像清华陈志华先生那样带着学生去考察，大家风餐露宿，半个月、一个月地在当地，我觉得现在的我根本干不了。

黄：我为什么提到考察，我来做一些回顾，20世纪上半叶，在中国近代建筑发展中，有许多建筑师对于传统建筑的喜爱与热衷远远超过现在的人，也许是那个年代里，中国社会正处于传统转向现代的过程中，所以相对当时距离中国古建筑比现在是接近了点。而我又观察到，中国近代建筑的调研与考察部分可以分为两种：一种是以正统价值观去考察，比如考察石阙、崖墓、佛窟、楼阁、寺庙、宫殿、木塔、桥等，这部分占了绝大多数；另一种是考察少数民族民居，比如考察村落，这部分占极少数。但后来陆续有建筑学者补充了这一块的内容，但对于偏远山区的村落与聚落，还是着墨很少。我想，这仍是当今中国建筑领域所存在的问题，我想问问您对这部分的看法？

朱：我们经常讲"继承传统"，但是继承传统，真的不是一件容易的事情。我们投入到"继承"这方面的精力太少了。我本人对"继承"的认知也停留在一个肤浅的层面上。传统建筑，如果你不进入到那个环境中去，不真心投入进去，你几乎没有任何心得。传统建筑考察，就像画画写生一样，你真好好去写生去画树，以后回头再画树，那个树就真的印在脑子里。因为你认真地写生过，你才知道松树的树干是怎样的走向，为什么在这里要拐一下、在那里要拧一下，转去又转过来，为什么老松树的下面盘根错节、乱七八糟，但是上面的部分要转折一下才能上去，你知道这就是松树的那种劲儿。如果你没有写生过，你永远不知道松树到底是什么样。

建筑更加复杂，你要不去了解、或你根本就没有做过传统建筑，你最好不要去碰传统的东西，一碰就要露馅儿。没有投入，不可能有产出。所以你看中国近代建筑师，他们这些人，真的是下了很大的工夫。像梁思成，他的父亲是梁启超，我觉得他们这一代人对皇家的东西、对正统的东西的研究下了工夫。但客观地说，中国传统建筑非常浩瀚，千差万别，皇家、官式的东西一定是在大量的民间建筑的基础上的一种提炼。梁先生的研究，大部分是关于皇家官式建筑的研究，但大量存在于民间的各种千差万别的东西，恰恰是中国传统建筑的核心部分，那就是因势利导、因地制宜。这种变化，才是我们中国传统建筑文化的精髓。传统建筑并非是固化的、形式上的东西。

黄：回到项目来，回到一个建筑史的视点，观察您的作品到后期的表述，可能会把您归到跟地域、传统稍微契合的这一条线上，那你自己怎么看待呢？虽然说您提到您的设计是强调建筑与环境的关系，但这只是您自己设计时的操作手法，若客观来看，不是每个项目都跟环境有关系，比如"秀"吧，看到的是宋式大屋顶，且因为它又放在一个非常现代时尚的躯壳上，以至非常吸引人，但若从建筑学的视角会观察到，这是一个强烈的传统语言的现代再现，然后空间的路径与场景也偏向于传统的轴线布局，用步移景异的方法塑造，这更会把您推向地域、传统的这一条线上，您自己怎么看待？

朱：传统文化的现代化发展，是一个永恒的题目，不仅建筑师要回答，任何一个行业、任何一个艺术门类都必须回答。传统如何走向现代，是我们当代人的责任，也是我的责任。我作为中国建筑师在国际设计竞争的氛围下，更应该主动承担起这样的责任。我们不能跟在别人后边盲目地跑。我到了这个年龄，更没有必要去尝试别人做过的事情，我还是愿意尝试、兑现一些我对传统的责任。你说的将我归为"地域、传统"这一类，我是认可的。我试图把传统的建筑文化通过现代的方式表现出来，而不是复古，希望能够对传统建筑文化的价值有新的尝试、新的诠释，在这一点上我一直在做。

你刚才提到的房子，实际上都跟传统文化元素有关。我不会把老房子照搬进来，或者直接将传统空间复制，而是考虑我的形式到底是什么。我能够一直把这个过程作为主线来研讨，后来我找到这个问题的关键点，那就是"放松"，怎么能够放松地来表达。

黄：您有没有一个自己设计的中心思想与信仰？问的比较形而上，就是您自己所追求的设计高点，是意境或者是境界，或是现阶段还处于摸索与摇摆的阶段。

朱：首先，我是一个真实的人，愿意以真实的一面来面对我所遇到的问题，所以希望能够回答设计的真相，而不是简单地去用我个人的设计方法。我做设计院的院长，也是同样的，有各种各样的方式可用来做掩饰，但我还是愿意追求真实的东西，这是做学问的学风，或者说，这是一个基本的态度。在这个前提下，我觉得所谓真实的东西，不是我个人主观的东西。建筑生成的因素，跟周围的环境有关，包括物理环境、自然环境、人文环境、历史环境，刚才谈了很多，这些东西都可能成为我做设计的参考系统。设计并不是简单的、个人的独断专行。

这个参考系统，我会认真对待，去分析建筑所处环境的特殊性。真正设计的过程中，我会在头脑中搜寻一种类型学的、解决问题的最基本的样式，这个样式可能是中国或外国曾经有过的解决方案。我会参照一些东西，比如院落，可能是从历史方面得来，可能通过院落关系找到了解决设计的问题与方法。既然强调历史跟人文的关系，我愿意从历史与现代的关系中找到他们之间的一个关联。比如，我把现代建筑寄希望于城市，建筑成为城市尺度扩大过程中的标记；现代建筑，由于其超大的尺度，从某种意义上讲，体现的是现代城市的关系，特别是在我们中国，跨街区的大项目，已经没有了建筑本身的意义，人来观察建筑时，已经没有建筑和人进行完整对话的可能性。所以，我将建筑定位在城市空间尺度的大的变化上，以这样的定位，找到一些环境的要素，来体会人和尺度的关系；在城市空间尺度的大的变化中，找到适宜人的尺度和环境。这是我设计的一个方法，或者说，一个理念。

作品年表
Chronology of Works

★——收录作品　●——实现作品　○——方案／在建作品

建筑名称	海南寰岛（泰得）大酒店 ●
主创建筑师	朱小地
项目所在地	海南省海口市海甸岛东部开发区
设计时间	1992年1月
竣工时间	1994年6月
占地面积	20000m²
建筑面积	46500m²
获奖情况	建设部部级优秀工程设计奖／建筑设计三等奖
	北京市第八届优秀工程设计／建筑设计二等奖

建筑名称	海南博鳌金海岸大酒店 ●
主创建筑师	朱小地
合作建筑师	杜　松
项目所在地	海南三亚
设计时间	1997年5月～1998年9月
竣工时间	1999年7月
占地面积	22368m²
获奖情况	第三届中国建筑学会建筑创作奖佳作奖
	全国第十届优秀工程设计奖／建筑设计银奖
	建设部部级优秀工程设计奖／建筑设计二等奖
	北京市第十届优秀工程设计／建筑设计一等奖

建筑名称	SOHO 现代城 ●
主创建筑师	朱小地
合作建筑师	贾更生、解 强
项目所在地	北京东长安街
设计时间	1999 年 12 月～ 2002 年 10 月
竣工时间	2003 年 05 月
占地面积	240000m²
获奖情况	建设部部级优秀工程设计奖 / 建筑设计三等奖
	北京市第八届优秀工程设计 / 建筑设计二等奖

建筑名称	盈创大厦 ●
主创建筑师	朱小地
合作建筑师	王 戈
项目所在地	北京西城区金融街
设计时间	2002 年 11 月～ 2005 年 8 月
竣工时间	2006 年 05 月
建筑面积	99218.29m²
占地面积	1.172ha

建筑名称	奥林匹克中心区规划与设计 ★
主创建筑师	朱小地
合作建筑师	张 果
项目所在地	北京
设计时间	2003 年 5 月
竣工时间	2008 年 7 月
占地面积	82hm²
获奖情况	2008 第五届中国建筑学会——建筑创作奖 (优秀奖)
	2009 年全国优秀工程勘察设计行业奖——建筑工程一等奖

建筑名称	西安唐大明宫国家遗址公园御道广场 ★
主创建筑师	朱小地
合作建筑师	樊则森、汪大炜
项目所在地	陕西西安
设计时间	2008 年 11 月
竣工时间	2010 年 9 月
占地面积	23.2 万 m²
获奖情况	2009 年全国经典建筑规划设计方案竞赛 规划金奖

建筑名称	中国石油天然气集团公司总部大厦 ★
主创建筑师	朱小地
合作建筑师	王 勇，吴 晨
顾问单位	英国TFP建筑事务所（方案）
项目所在地	北京
设计时间	2001年11月~2002年05月
竣工时间	2008年8月
占地面积	143000m²
建筑面积	200838m²
获奖情况	2009年住房和城乡建设部科技示范工程
	住房和城乡建设部建筑能效测评等级证书（三星级）
	美国国际绿色建筑Leed金奖
	2011年全国优秀工程勘察设计行业奖——建筑工程一等奖

建筑名称	"山水楼台"会所 ★
主创建筑师	朱小地
合作建筑师	宓 宁
项目所在地	北京怀柔
设计时间	2002年4月
竣工时间	2003年8月
占地面积	2219.82m²
建筑面积	5500m²

建筑名称	银泰中心南裙房屋顶花园（"秀"吧）★
主创建筑师	朱小地
合作建筑师	宓 宁，钟 菲，杨 波
项目所在地	北京银泰中心南裙房屋顶
设计时间	2006年6月
竣工时间	2009年4月
建筑面积	1300m²
获奖情况	2011年全国优秀工程勘察设计行业奖——建筑工程二等奖

建筑名称	哈德门饭店 ★○
主创建筑师	朱小地
项目所在地	北京崇文门
设计时间	2006年6月
建筑面积	160602m²

建筑名称	"旬"会所 ★
主创建筑师	朱小地
合作建筑师	高　博
项目所在地	北京东四环百子湾路 21 号院
设计时间	2007 年 8 月
竣工时间	2010 年 8 月
建筑面积	1664m²

建筑名称	"青"会所 ○
主创建筑师	朱小地
合作建筑师	金国红、陈　莹
项目所在地	北京市朝阳区
设计时间	2010 年 7 月～2011 年 2 月
占地面积	4000m²
建筑面积	2263m²

建筑名称	深圳第一创业大厦 ○
主创建筑师	朱小地
合作建筑师	罗　靖、刘昕欣
项目所在地	深圳市中心区
设计时间	2008 年 10 月
占地面积	4111.12m²
建筑面积	50551m²

建筑名称	全国组织干部学院 ●
主创建筑师	朱小地
合作建筑师	叶依谦
项目所在地	北京市朝阳区
设计时间	2009 年 5 月
竣工时间	2011 年 4 月
占地面积	163440.2m²
建筑面积	40062m²
获奖情况	住房和城乡建设部三星级绿色建筑设计标识证书

建筑名称	平安里地铁站后续工程 ○
主创建筑师	朱小地
项目所在地	北京市新街口南大街
设计时间	2011 年 7 月
占地面积	8725m²
建筑面积	15000m²

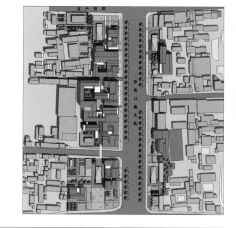

建筑名称	珠海歌剧院 ○
主创建筑师	朱小地
合作建筑师	马 泷、栾 波
项目所在地	广东珠海
原竞赛合作单位	陈可石 (北京大学中国城市设计研究中心)
设计时间	2009 年底至今
占地面积	57670m²
建筑面积	59000m²

建筑名称	嘉莲会所 ○
主创建筑师	朱小地
合作建筑师	林 卫、徐通达
项目所在地	陕西西安
设计时间	2011 年 6 月
竣工时间	2012 年 4 月
占地面积	1730m²
建筑面积	4540m²

建筑名称	平安里地铁站后续工程
主创建筑师	朱小地
项目所在地	北京市新街口南大街

朱小地简介

1964年5月生于北京
1988年毕业于清华大学建筑系建筑学专业
北京市建筑设计研究院院长、总建筑师

教授级高级工程师
国家一级注册建筑师
政府特殊津贴专家

代表作品

海南寰岛（泰得）大酒店、海南博鳌金海岸大酒店、哈德门饭店、光大花园规划及5号、6号、7号住宅楼设计、观湖国际住宅区、万科西山庭院规划、北京SOHO现代城、盈创大厦、中国石油天然气集团公司总部大厦、深圳文化中心、西安唐大明宫国家遗址公园御道广场、奥林匹克公园中心区规划与设计、银泰中心南裙房屋顶花园（"秀"吧）、北京"旬"会所等。

主要获奖项目

海南寰岛（泰得）大酒店
1998年建设部部级优秀工程设计奖——建筑设计三等奖

海南博鳌金海岸大酒店
2002年全国第十届优秀工程设计奖——建筑设计银奖
建设部优秀工程设计二等奖

北京SOHO现代城
2002年全国第十届优秀工程设计奖——建筑设计铜奖
2002年建设部优秀工程设计二等奖
2004年第三届中国建筑学会建筑创作奖（优秀奖）

奥林匹克公园中心区规划与设计
2009年全国优秀工程勘察设计行业奖——建筑工程一等奖
2008年第五届中国建筑学会建筑创作奖（优秀奖）

深圳文化中心
2009年全国优秀工程勘察设计行业奖——建筑工程一等奖（合作设计）

中国石油天然气集团公司总部大厦
2009年住房和城乡建设部科技示范工程
住房和城乡建设部建筑能效测评等级证书（三星级）
美国国际绿色建筑LEED金奖
2011年全国优秀工程勘察设计行业奖——建筑工程一等奖

银泰中心南裙房屋顶花园（"秀"吧）
2011年全国优秀工程勘察设计行业奖——建筑工程二等奖

全国组织干部学院
住房和城乡建设部三星级绿色建筑设计标识证书

西安唐大明宫国家遗址公园御道广场
2009年全国经典建筑规划设计方案竞赛 规划金奖

Profile

Born in May 1964, in Beijing
Graduated from the architectural department of Tsinghua University in 1988
President and Chief Architect of Beijing Institute of Architectural Design at presentd

Professor-grade senior architect
National Registered Class-1 architect
Expert subsidized with special government allowance

Representative Projects

Huandao Hotel in Hainan Province, Boao Golden Beach Hotel in Hainan Province, Hademen Hotel in Beijing, planning of Guangda Garden and design of Building 5, 6 and 7, Guanhu International Residential Quarters, Vanke Xishan Court planning, Beijing SOHO Modern City, Yingchuang Building, China Petroleum Headquarter Building, Shenzhen Culture Center, Emperor's Way Square of Xi' an Daming Palace National Heritage Park, planning and design of the Olympic Park, roof garden on the South Podium of Yintai Plaza (Xiu Bar),Xun Club.

Awards

Huandao Hotel in Hainan Province
The 3rd Prize of Architectural Design of Excellent Construction Design of the Ministry of Construction, 1998

Boao Golden Beach Hotel in Hainan Province
Architectural Design Silver Prize of 10th China Excellent Construction Design
The 2nd Prize of Excellent Construction Design of the Ministry of Construction

Beijing SOHO Modern City
The 10th China Excellent Construction Design Award 2002 - Bronze Award of Architecture
The 2nd Prize of Excellent Construction Design of the Ministry of Construction, 2002
Excellent Prize of the 3rd Architectural Innovation Award of Architectural Society of China, 2004

Planning and design of the Olympic Park
The 5th Architectural Innovation Award of Architectural Society of China, 2008
China Excellent Engineering Survey and Design Award 2009, the 1st Award of Architectural Construction

Shenzhen Culture Center
China Excellent Engineering Survey and Design Award 2009, the 1st Award of Architectural Construction (cooperative deisgn)

China Petroleum Headquarter Building
Technological Sample Project 2009 of the Ministry of Housing and Urban-Rural Development
Energy Efficiency Grade Certificate (3-star) of the Ministry of Housing and Urban-Rural Development
LEED Golden Prize of the International Green Building, the United States
China Excellent Prize of Construction Survey and Design 2011, the 1st Award of Architecture

Roof Garden on the South Podium of Yintai Plaza (Xiu Bar)
China Excellent Engineering Survey and Design Award 2011, the 2nd Award of Architectural Construction

China Organization and Cadres Institute
3-star Grade Certificate of Green Architectural Design awarded by the Ministry of Housing and Urban-Rural Development

Emperor's Way Square of Xi' an Daming Palace National Heritage Park
Golden Prize of City Planning of China Building Planning Design Competition 2009